IMPORTANCE OF SEVERAL MATHEMATICAL REASONING

Edward Grant, PhD

Copyright © 2021 by Edward Grant

All rights reserved. No part of this publication may be reproduced, stored in a retrieval system, or transmitted in any form or by any means, electronic, mechanical, photocopying, recording or otherwise, without the prior permission of the publishers.

PREFACE

Mathematics is an exact, hard science. Mathematical ideas have been used in all fields of work, e.g., engineering, computers, commerce, etc. This book touches on both applied and pure mathematics. In applied mathematics, the Navier-Stokes equations have been used to model and forecast fluid flows, which are important in the fields of aeronautical and marine designs. In pure mathematics, the prime numbers are important in cryptography and computer security work.

This book would expose readers to such important mathematical reasoning.

Edward Grant, PhD

CONTENTS

1 Why Mathematics?   5
2 Logic In Mathematics   6
3 Application Of Navier-Stokes Equations   20
4 Navier-Stokes Equations And Limitation   23
5 Calculus And Differential Equations   43
6 Primes, Twin Primes And So On   46
7 Interesting Twin Primes   59
8 Primes And Even Numbers   65
9 Primes Distribution And Riemann Hypothesis   94
  Further Reading   97

1

WHY MATHEMATCS?

Mathematics is the language of symbols. Mathematical symbols are used to compress information in a way that no other language possibly could.

Mathematics as a language may take various forms, e.g., arithmetic, algebra, geometry, statistics, etc. It could be divided into two broad branches, viz., pure mathematics and applied mathematics. Pure mathematics is studied for its intrinsic aesthetic appeal while applied mathematics is mathematics which is used in our practical work, e.g., in engineering and commerce. It may come as a surprise to some, though not to the professional mathematician, that some ideas in pure mathematics do have some applications in our everyday affairs, e.g., in scientific work wherein the phenomena of nature have to be interpreted or explained. Many scientists think that nature is mathematical, and, as such, mathematics has been utilised to model some aspects of nature.

Mathematics may be beautiful to those who are able to appreciate its quality but is awe-inspiring and even irksome to many others. Why is this so? It is perhaps due to the obscurity of the symbols used and poor mathematical teaching in schools that are the cause of this unbecoming state of affairs.

But it is considered a hard science, an exact science, though some of the more abstruse aspects may be controversial and may thus be hardly classifiable as "exact". The objects of mathematics, viz., numbers and quantities, may be regarded as "exact", but the many ways and methods of dealing with these objects in mathematics may hardly be "exact" and may even be controversial.

The big question here is whether mathematics is discovered or invented. In other words, is mathematics part of nature which we have discovered or which is waiting to be discovered. It seems to be a bit of both - invented as well as discovered - though some mathematicians may prefer to regard mathematics as a reality independent of human existence, a reality which human beings will sooner or later discover.

Mathematics as a hard science which apparently demands much from our logical faculty may seem unimportant to the non-mathematician or the non-mathematical, but its importance in human affairs cannot be denied. For instance, without the mathematical procedures known by the term "algorithms", our computers would not be able to carry out any computation. Other examples of the usefulness of mathematics are the use of statistics by the insurance industry and the use of operations research techniques in the manufacturing industry, etc. A more general use for mathematics may be that it is a means of sharpening the mind's capacity for logical thinking.

2

LOGIC IN MATHEMATICS

Say, we are faced with several premises or statements. How shall we proceed from here to a truly logical conclusion? How do we proceed hence with the act of logical reasoning? How we wish logical reasoning could proceed so smoothly and surely as "1 + 1 = 2" or "1 + 1 + 1 = 3" or "A is taller than B. B is taller than C. Therefore, A is taller than C." or "A is taller than B. B is taller than C. D is shorter than B but taller than C. Therefore, A is taller than C and D." Notice that such "conclusions", which are the result of the act of logical reasoning, are measurable, quantifiable or verifiable entities (in this case, quantity and height respectively). But, unfortunately, in many cases the conclusions derived from the act of logical reasoning is not measurable, quantifiable or verifiable, the more abstract and less "practical" or "practicable" the conclusions, the less measurable, quantifiable or verifiable they would be. (This would be, as stated earlier, a case of "quantitative" logic versus "qualitative" logic.) This is indeed the greatest problem with logical reasoning. Since a conclusion is not based on concrete, tangible or verifiable facts, or, premises, there is no way to tangibly or physically measure, check, evaluate or verify the conclusion. At most, we could have a very strong intuitive feeling or "gut feeling" that the conclusion based on those premises is correct or logical.

In logic, "seeing is believing" must be everything. What we could not see, measure or physically verify must be subject to some doubt. There is no point in insisting and swearing that one's conclusion is absolutely correct when there is no "physical basis" or "tangible, verifiable basis" for one's so-called logical conclusion. Where the conclusions are not measurable or verifiable, the best recourse should be to regard the conclusion as <u>tentatively</u> correct or logical; the best, most practical and most logical stance is to wait and see, compromise, or give and take. Otherwise, we might end up with "unconvincing" arguments, circular reasoning or confusion.

The author would like to propose the following ways of going about the act of logical reasoning:-

1) List all the possible implications of the premises (or statements).
2) Find and list the logical links between these implications, if any.
3) Form a logical conclusion from these logical links.
4) Finally, and very importantly, attempt to <u>physically verify</u> the conclusion, if this at all possible. (Just like the scientist carrying out a physical experiment to verify a hypothesis.) At the very least, try to find out whether there was or were any parallel(s) in the past, i.e., verify whether any similar happenings or "conclusions" have taken place in the past as a result of similar circumstances or "premises". (Carry out thorough research work.)

Once again, the author stresses that logical reasoning should be, as much as possible, based on solid, tangible, measurable, quantifiable, verifiable facts in order to be "trouble-free", truly correct or indubitable. If the logical reasoning is based on abstract premises alone it is likely to be hard to follow or understand and to cause doubt. It is when logical reasoning is based on solid, tangible and verifiable facts, that it is easily comprehended, clear and indubitable. When this happens, disputes would be minimal, and there would be comparative harmony and peace. In short, "quantitative" logic, whose results are measurable, verifiable, is easier to handle than "qualitative" logic, whose results are not tangible or verifiable and might be subjective.

One additional bit on the use of logic here. Mathematicians, who are theoretically well-trained in the use of logical reasoning, have been attempting to apply logical reasoning to solve mathematical problems pertaining to infinity, e.g., Euclid's proof of the infinity of the primes, the twin primes conjecture pertaining to the infinity of the twin primes, the Goldbach conjecture pertaining to the infinity of the even numbers being each the sum of two primes and the Riemann Hypothesis pertaining to the infinity of the solutions of the Riemann Zeta function which lie on a straight line which would imply the infinity of the primes and describe their distribution. It is apparent that to the mathematicians logical reasoning is a very powerful tool for solving very difficult problems such as those mentioned above. Now, could they or any logician not use the same kind of rigorous logical deduction to prove the existence of God (or, even the existence of the Devil or Satan), just as the mathematician, Descartes, had tried to use logical deduction to prove that God exists? In actuality the concept of infinity is rather "mind-blowing". No one could count, check, verify or live to infinity. It is just an abstraction, a concept. We could conceptualize the idea of a Being that lives to infinity or eternity and that would be the idea of God. If a mathematician could prove the infinity of some mathematical objects such as the infinity of the twin primes or the even numbers which are each the sum of two primes, they could in theory prove, or disprove, the possible existence of a Being who lives for infinity or eternity, i.e., God, who could be regarded as the embodiment of infinity or Infinity itself personified. The mathematician, Georg Cantor, had proved that there are different orders or degrees of infinity (by using a "diagonal" method), an idea so bizarre that he had been viciously criticized by his contemporaries, which had led to his mental breakdowns, though this concept of infinity is now embraced by mathematicians. In other words, some infinities are larger than others. All this is a question of interpretation, of how one chooses to interpret infinity, an interpretation which is arbitrary and not subject to any definite and clear-cut rules. To the author's interpretation, an infinite sequence or progression is some object that goes on forever (with no ending), and there has never been a higher or greater, or, lower or lesser degree or magnitude of "forever". Mathematicians in effect tell us now that there are different degrees or magnitudes of "forever". Nobody has ever directly experienced infinity, something that goes on forever, and suddenly we are told that there are infinities which are larger than others. To practically all of us there is only one "forever", one infinity. It is as though we are hearing someone tell us that one corpse is more dead than another, or, this one inch is longer (or shorter) than the other one inch, or, worse, some one inches are longer (or shorter) than other one inches. How ridiculous it all seems. (Here, it may be appropriate to elaborate a little further on this question of the different degrees of infinity as postulated by Cantor. Cantor had used a very clever technique, which could perhaps be viewed as a cunning "trick" if one may regard it as such. He had listed rows/columns of various numbers, the numbers representing members of the various sets. Then, by the "diagonal" method he had created new sets whose members were distinct from the members of the original list. Cantor had interpreted this phenomenon as the existence of different infinite sets with different degrees, magnitudes or orders of infinity. But, all these various sets of numbers could perhaps be viewed en masse as one set of numbers which are "infinite in all directions" - horizontally, vertically, diagonally, zigzag even, in fact any other directions, i.e., there is only one infinity, and not various infinities. If we, e.g., substitute the "quantity", infinity, here with the "quantity", one inch, we could now consider ourselves having an one inch (which is actually equal to 25 mm) with different degrees or orders of magnitude. We could then consider ourselves having, e.g., an one inch which is seven mm, one which is eight mm, one which is nine mm, one which is ten mm, and so on, etc., i.e., we have one inches of various degrees of magnitude, which are the analogues of Cantor's infinities with different degrees or magnitudes. Is this not ridiculous?) Thus, it appears, logic is subject to different interpretations and rules, and, hence, its resultant subjectivity, and problems.

Earlier, it has already been stated that one's capacity for logical thought is a function of one's ability to use a certain medium (in this case, a language) for carrying out the act of logical thinking. Thoughts and ideas, logical thoughts and ideas, are conceived in words (which could be read), sounds (which could be heard) or symbols (which could be seen/heard and interpreted). Mathematicians are experts in the use of symbols, mathematical symbols to be precise. Others who make use of logic, such as, e.g., lawyers, make use of the medium of communication, e.g., the English language. Therefore, one's mastery of a medium of communication, a language in which to carry out logical deduction, is of very great importance. The English language, in all its complexities and nuances, is a very important medium for the performance of the act of logical deduction, despite its disadvantage of being able to lead to misinterpretation of meanings. The English language, it should be noted, is able to convey the subtlety of ideas or logical concepts, and is often subject to abusive usage by those who like to "twist and turn" with ideas or arguments. The author however is of the opinion that the English language, being an international language and a most popular and universal one too, is a most effective tool for logical deduction; its mastery is of paramount importance for logical deduction; even in mathematics, the English language is of fundamental importance. Mathematics may rely on a copious use of symbols (signs which have meanings), which are a form of "short-hand" in which to carry out mathematical reasoning; in other words, the mathematical symbols facilitate mathematical reasoning. But each mathematical symbol, which represents an idea or thought, would have to be interpreted by the human brain in a language with which it is familiar, a medium of communication, e.g., the English language, Greek, French, German, etc. And, the language itself, be it English, Greek, French, German, or any other, is the set of "symbols" which represent the objects which could be experienced by our senses, viz., sight, hearing, smell, taste and touch. The mathematical symbols could be equated with the words of any language or medium of communication. An examination of any mathematical journals or texts would reveal that mathematical symbols alone are insufficient and are normally accompanied by words, the language of communication, e.g., English. It is apparent that most mathematics journals are published in English. Someone not familiar with the symbols in the mathematical text though he might be an expert in the English language would tend to find the text rather incomprehensible, which is not surprising. For example, the classic two-volume mathematical tome, Principia Mathematica, which had been co-authored by Bertrand Russell and Alfred N. Whitehead and which brought fame to its authors, are fully embellished with symbols and practically devoid of the normal language and are by all accounts incomprehensible even to the keenest mathematicians, though the ideas in the tome, which are incomprehensible because of the arcane symbols used to convey them, are not that difficult. The author is of the opinion that the mastery of the language of communication, e.g., English, is more important than the mastery of the mathematical symbols (which could be translated into, or has to be understood in terms of, the language of communication, e.g., English). Even the mathematical symbols are normally created from the letters of a language of communication, e.g., English and Greek, something which is familiar. This has to be the case for, as stated earlier, ideas and thoughts could not exist without the meaningful words or media of communication with which our minds are familiar. It is practically impossible to think without language or words, and, therefore, the better one's mastery of a language is, e.g., the English language, the more capable one is of logical deduction. We may make use of certain images or signs and may even communicate with signs, but our thoughts are in the terms of a language, or, words, with which we are familiar. All the seemingly abstruse symbols in mathematics could be translated into plain English words or the words of any other language, and, vice versa.

Many people fear mathematics and have problems understanding mathematical ideas, especially higher mathematics with its arcane symbols (which may also have the dual objective of keeping "outsiders" out or in the dark). The problem is due to the obscurity or ambiguity of the mathematical symbols themselves. It is because we are uncertain what the mathematical symbols represent and are unsure of how to interpret these symbols that make us want to throw up our hands in trying to follow or understand a chain of mathematical reasoning, especially so if the chain is a long and convoluted one (a good memory would be of great help in order to grasp with ease the long chain of reasoning involving plenty of abstruse symbols and unfamiliar terms - hence, the importance of a good memory to logical deduction). However, these "frightful" mathematical symbols could be translated into the plain, simple words of a language with which we are familiar, and every mathematical reasoning could be as clear as daylight. Therefore, in the very act of logical deduction, we should avoid obscurity and ambiguity and make our ideas or reasoning as clear, precise and simple as possible. The author believes that no idea or logical reasoning is too difficult to be understood. Where the idea or logical reasoning seems too difficult to comprehend it is most likely to be the fault of the act of exposition or the use of the wrong words or symbols. Take the following example; if the author were to use the following "abstruse" symbols to convey a message which is in actual fact very simple, how could anyone fathom what he is saying?:

$$\Xi \quad \nabla \quad \varpi \approx \zeta \quad \phi \quad \Theta \quad \varnothing \quad \lambda \not\subset \alpha$$

Perhaps, Sherlock Holmes could have deciphered this message.

In life, in politics, practically everyone would favor democracy. We want to have the right to vote, the right to express our views and be heard, the right to live freely, etc. Why not extend these democratic principles to our intellectual life as well? We preach equality when we espouse democracy. Whether rich or poor, Dutch or English, Christian or Buddhist, etc., we want to be treated equally. Then we should treat everyone as our intellectual equals as well and give everyone a chance to express their views and thoughts and respect their views and thoughts, instead of castigating others for holding certain views and subjecting them to ridicule or showing disrespect or discourtesy to them for having those views. In the use of logic, we should be democratic enough to listen, respect, share and compromise.

It is the habit of experts or professionals in various specializations to use jargon, terms or expressions peculiar to their specializations when communicating. For example, lawyers are fond of legal terms or jargon, what we call "legalese", engineers are prone to using their own technical jargon or technical terms and mathematicians or logicians have their own special symbols and terms. It is supposed that this is to add a touch of "professionalism" or "image of special knowledge" for these various specialists, and the uninitiated who is not privy to their specializations is not supposed to understand what they are saying and so would be kept out - this could be regarded as a form of "knowledge protectionism". This could also be made out in such a way that these specialists would appear more profound than what they actually are. In fact, some philosophers have been known to have been accused of couching their ideas, which might indeed be very simple, in high-sounding or profound-sounding terms so as to be impressive. Is all this bad? All this is likely to lead to lack of understanding (here, the ideas serve no purpose - if nobody understands them they are as good as not having been said - in fact it is a waste of the target audience's time to utter these ideas if the philosopher knows that these ideas are going to be incomprehensible - the philosopher may just want to appear profound and impress but this is selfish and should be discouraged), misunderstanding (which is worse than lack of understanding as it may cause problems), confusion, or even conflict, on the part of the uninitiated. It is as though the specialists are keeping out the uninitiated. Therefore, in the exercise or use of logic it is proper, and wise, to avoid incomprehensible or difficult to comprehend jargon or specialized terms, and utilize clear and simple terms in order to have the logical connections easily and clearly understood, and for there to be no confusion. The important aim of logic should be to convince or convert and not to confuse. Unfortunately, this is often not the case. Here, the author emphasizes that logic should rely on simple, clear, concise and precise terms and expressions as far as possible. The English language, e.g., is capable of meeting this requirement but, unfortunately, its usage is often abused, with untoward consequences.

To the author, logic could be regarded as the beauty of ideas or concepts, which is not unlike physical beauty. Physical beauty is subjective and qualitative, and not quantitative, but at least it describes a tangible person or object. On the other hand, the beauty of ideas, though also subjective and qualitative, describes something intangible (not physical or concrete) or abstract. As logic, like physical beauty, is thus subjective, all the more it should always be held with suspicion or even skepticism. We should never force it on others as others might have their own standards of logic, their own standards of beauty for ideas. (As we normally do not quarrel about or dispute over physical beauty, claiming that it is only "in the eyes of the beholder", so similarly should we regard logic, or, the beauty of ideas as such.) Logic could be regarded as an invention of the human mind which could not exist independently of the human mind, unlike, say, the existence of tangible objects such as rocks, trees or water. It could regarded as merely a play of words or language (as the great philosopher, Wittgenstein, had postulated, for, also, as mentioned earlier, it is not possible to conceive ideas, logic, without words or symbols; could a blank but conscious mind think logic, could one carry out the act of logical deduction without words or symbols?)

Logic could even be a tautology, describing an object in different terms but actually conveying the same thing, e.g.:-

> It has long hair and wears a skirt and high heels.
> It is not a man.
> Therefore, it is a woman.

In any case, even without the above logical deduction, we all know that if a person is not a man, he is, he must be, a woman (though there are very rare cases of hermaphrodites or persons with both male and female features), i.e., the negative of man is woman. In other words, "woman" has the same meaning (synonym) as "not a man" - a tautology so to speak, using different expressions but meaning the same thing.

The author would like to bring up some examples of logic being subjective and merely some rules invented by the human mind for arriving at some logical conclusions. These examples are taken from mathematics, where logic is supposed to figure prominently, where its practitioners take pride in the rigor of its logic. As the reader is probably aware, mathematical reasoning relies much on certain axioms or assumptions, which are obvious truths that do not require proofs (the problem here is what is obviously true to one person might not be obviously true to another), lemmas and theorems (the problem of lemmas and theorems, which are truths which have proofs, is somewhat similar to that for axioms in that what are considered as the proofs of the truths' validity might not be acceptable or obvious to some others). Without them no mathematical reasoning might have been possible. Of course, if they were wrong, then the results of the mathematical reasoning would have been wrong, and meaningless. This seems unique only to mathematics, for no other types of reasoning, e.g., scientific, legal and economic reasoning, work in such a manner, i.e., rely on axioms, lemmas and theorems, though they might make assumptions (which could be considered the equivalent of axioms). Since mathematical logic is reputed for its rigor, should not other types of reasoning, like the three mentioned above, proceed ahead with their respective axioms, lemmas and theorems, if this were at all possible and practical? On the other hand, why couldn't mathematical reasoning be more mundane and more like our normal, usual way of reasoning? Is our normal, everyday mode of reasoning, though it might have helped produce great results and has brought great problems as well really inferior to mathematical reasoning? The author has had some doubts about some of the so-called mathematical proofs or reasoning, which seem rather artificial, arbitrary and lacking in rigor.

Consider the following mathematical theorem:-

If α and β are cardinal numbers (finite or infinite):

$$\alpha \leq \beta, \ \beta \leq \alpha \rightarrow \alpha = \beta$$

Stated in everyday language, if alpha is smaller than or equal to beta, and, beta is smaller than or equal to alpha, then alpha is equal to beta. But, there seems to be a contradiction here - on the one hand it is stated that alpha is smaller than (or equal to) beta, and, somewhat contrariwise, it is stated that beta is smaller than (or equal to) alpha, which casts some doubt on the credibility of the logic of the conclusion that alpha is equal to beta, there being not quite a logical link between the premises, "alpha is smaller than or equal to beta", and, "beta is smaller than or equal to alpha", and, instead, there appears to be a contradiction between them - "alpha is smaller than beta" contradicts "beta is smaller than alpha". There appears to be some other way of interpreting, explaining or proving this theorem, but the author has no knowledge of it - this theorem has been proven independently by E. Schroder (1896) and F. Bernstein (1898), and is known as the Schroder-Bernstein Theorem. To the author, we should not definitely conclude from the two statements or premises, alpha is smaller than or equal to beta, and, beta is smaller than or equal to alpha ($\alpha \leq \beta$, &, $\beta \leq \alpha$) that alpha is equal to beta ($\alpha = \beta$); indeed, three possibilities could be deduced from these two statements or premises: either alpha is smaller than beta ($\alpha < \beta$), or, beta is smaller than alpha ($\beta < \alpha$), or, alpha is equal to beta ($\alpha = \beta$) - the probability of each of these three possibilities being realized being 33.33%. On the other hand, if alpha is an integer or whole number smaller than 2, and, beta is also an integer smaller than 2, then we could certainly conclude that alpha is equal to beta, that is: $\alpha < 2$, $\beta < 2$ $\rightarrow \alpha = \beta$ - in this case, alpha is equal to 1 and beta is also equal to 1, wherein alpha is obviously equal to beta, that is: $\alpha = 1 = \beta$.

Let us look at the following proof or mathematical reasoning:-

Proposition: If A, B, C are sets with $A \subseteq B$ and $B \subseteq C$, then $A \subseteq C$.

Stated in everyday language, this proposition stipulates that if A, B and C are sets, with A being a subset of B and B being a subset of C, i.e., every element of A is an element of B and every element of B is an element of C, then A is a subset (or element) of C.

Proof: Every element of A is an element of B and every element of B is an element of C. Therefore every element of A is an element of C, so $A \subseteq C$.

It appears that the above proof is merely a re-statement, or, repetition of the proposition, and seems hardly a proof though it is regarded as one. ($A \subseteq C$ could be read as or interpreted as "A is a subset of C wherein every element of A is an element of C".)
The following mathematical reasoning might appear questionable:-

Let $P_{(a)}$, ..., $P_{(n)}$ be a sequence.

Then, by induction, this sequence implies the presence of other sequences, as shown below:

$P_{(a)}$, ..., $P_{(n)}$ $\rightarrow$ $P_{(b)}$, ..., $P_{(n+1)}$
$P_{(b)}$, ..., $P_{(n+1)}$ $\rightarrow$ $P_{(c)}$, ..., $P_{(n+2)}$
....

The arrow ($\rightarrow$) represents "leads to". If we, e.g., let $_a$ above equal 1 ($_b$ is one step after $_a$ and $_c$ is one step after $_b$, as in the alphabet) and $_n$ above equal 10, then the above mathematical statements could be translated into the following:

1, ..., 10 $\rightarrow$ 2, ..., 11
2, ..., 11 $\rightarrow$ 3, ..., 12
....

In other words, induction tells us that if we see 1 ($P_{(a)}$), we could expect to see 2 ($P_{(b)}$) next, and if we see 2 ($P_{(b)}$), we could expect to see 3 ($P_{(c)}$) next, and so on. Induction is commonly used in mathematical proofs or mathematical reasoning. The question is whether its results are justifiable or rigorous. If, for instance, we see seagulls appearing at the beach every summer, we naturally conclude that seagulls would be at the beach every summer. This is induction, whereby a pattern or trend is observed. Is induction fool-proof? Definitely not. In fact, some mathematicians frown on induction as a method of mathematical reasoning. There is no solid reason why seagulls must appear at the beach every summer and it might not be logical, and bc too much, to expect seagulls to appear at the beach in the future summers. How could we be certain that the seagulls would turn up at the beach again in the future summers; is it not possible for them to find a better beach to frequent by then?

In mathematical reasoning, axioms or obvious truths which are not proven are frequently used. The author has some doubt about the use of axioms in mathematical reasoning. Mathematicians require "airtight" logic or rigorous proofs in solutions to mathematical problems and yet could accept the use of axioms. To the author, this is a contradiction. What is an axiom or obvious truth to one person might not appear so to another. Earlier too, the author mentioned some well-known mathematicians who needed 100 pages or so to prove such an obvious truth as: $1 + 1 = 2$, which could have been simply treated as an axiom. Is this not too much? It is as though one is trying to split hair, which is quite impossible and absurd. The author understands that axioms often have to be used before the mathematical reasoning could proceed for they act as the starting points or bases. But, at least they should be used with discretion, and, qualification, e.g., we could state that the axioms might be subject to doubt by some quarters but they could for feasibility be regarded as true, or, they could be regarded as provisionally true till proven otherwise. The imposition of this condition makes the reasoning all the more reasonable, all the more logical.

Also, mathematicians frequently resort to the "reductio ad absurdum" proof or proof by contradiction. It has been mentioned earlier and would be emphasized here that this kind of proof has a weakness. Here, the mathematician makes an assumption that a conjecture is wrong, or, correct, and then goes on to find evidence which contradicts this assumption, proving otherwise. For example, if an assumption that a conjecture is wrong is proven incorrect:

1) It might not necessarily mean that the conjecture is correct. (It could be partially correct, or, correct under certain circumstances and wrong under other circumstances. It could be half-correct and half-wrong. Some things could neither be classified as correct or wrong, e.g., an act might not be based on a moral or correct, or, immoral or wrong principle - it might be amoral. It could be just "average" or "in-between correct and wrong".)
2) Its correctness or wrongness might be un-decidable, i.e., it is neither possible to prove its correctness nor wrongness. (Reference could be made to Godel's Incompleteness Theorems.)

As many mathematical proofs are proofs by contradiction, there is the implication that there is a flaw or defect in much of mathematical logic. Godel's Incompleteness Theorems testify this - there is a serious defect in the foundation of mathematics. Hence, mathematical logic appears flawed in many aspects.

Though mathematics might be a comparatively difficult subject to master, it nevertheless plays a very important role in our lives. We need mathematics for calculations and computations, e.g., in engineering and computer work. Mathematics uses equations and formulae, which are comprised of symbols and letters, as well as the operations of addition, subtraction, multiplication and division. It deals with abstract concepts such as equality, sizes, degrees, elements, sets, groups, convergence, divergence and infinity, and, this is perhaps where the difficulty really lies, especially when the abstraction of mathematical concepts is carried out to such an extent that the ideas appear to be detached from reality.

Pure mathematics, like chess, might be an intellectually demanding art with sets of rules to follow, which has little utility or practical value. Even a mathematician and a real genius with incredible powers of comprehension such John von Neumann had difficulty with mathematics. Here we quote him, "One never understands mathematics, one simply gets used to it". In pure mathematics there are not only many rules one has to be familiar with but one has to get used to the many symbols and terms as well. Learning higher mathematics is not unlike learning a difficult foreign language. But the author thinks that learning a truly difficult foreign language is more difficult. What is the point of learning mathematics then if it is difficult and has little utility or practical value? The renowned British mathematician, G. H. Hardy, had lamented that as a pure mathematician by profession his contribution to society had been practically nil.

The only utility perhaps is that pure mathematics and chess, like the beautiful paintings of an artist, may bring joy and happiness to those who practice them or deal with them. However, some areas of pure mathematics, such as group theory and topology, e.g., have found practical uses in such fields as physics and biology. As to logic itself, the author would like to subscribe to the "good salesman" concept. Though logic may bring joy to those who appreciate its "beauty", it is in effect no better off than beautiful pieces of art, if it does not enable us to solve problems and progress. Good and effective logic is comparable to a good and effective salesman. What are the criteria for being a good and effective salesman? Is a good salesman a salesman who just carries himself well and is capable of making superb sales presentations? If you were an employer who has to choose between firing one of your two salesmen, one of whom is very impressive and capable of superb sales presentations but produces poor sales results, while the other salesman is not impressive at all and not so capable of making good sales presentations but produces excellent sales results, who would you fire? Wouldn't it be the former? The same concept should apply to logic as well. Logic should not just appear nice and impressive and bring happiness or pleasure. It should be capable of bringing concrete results, it should be capable of being "creative" or productive, it should be result-oriented. Like the beauty of art, where they say that "beauty is in the eyes of the beholder", the validity or truth of logic is "in the mind of the thinker". As the saying goes, "One man's meat is another man's poison". Perhaps, we could state, "One man's logic is another man's nonsense".

We know that logic is imperfect and creates problems for practically everyone. The author has along been trying very hard indeed to find a way out of this rut. Since logic often could not convince us as to what is right and what is wrong, we need a third party who is neutral and impartial to decide for us what the right paths are, a third party, whether human or artificial intelligence, who is sanctioned by the law, the authorities, to guide us through the right and logical paths. We need a highly intelligent computer, which is capable of storing vast quantities of data and quick and efficient computation to assist us, to guide us, a powerful thinking or logic-analysing machine which would indeed be impartial and neutral; all we have to carry out is to input our requests to this super-computer and the answers or advice would spew forth expeditious, efficiently and impartially from it. This is a distinct possibility and an apparently ideal solution pertaining to our perhaps overwhelming problems with logic. When this happens, if it does happen, there could be lasting harmony and peace. And, we could all sit back and sigh, "Peace at last!"

But, before we have the luxury of such a thinking, logic-analysing machine, we have to have an efficient and proper way of handling logic. We pride ourselves in our intelligence, our ability to understand and to reason (i.e., use logic). For example, the engineer uses logic (including mathematical reasoning) to analyse and solve engineering problems, the scientist relies heavily on reasoning or logic in analysing natural phenomena, the politician has to understand the needs of his constituents and has to find solutions to their needs, which requires logical reasoning, the chef has to figure out what food ingredients to use and in what proportions to arrive at dishes which stimulate the taste-buds and to some extent this requires logical reasoning, even the house-wife has to figure out how to spend on consumables within a given budget, an act which requires some degree of logical reasoning, etc. We see that logic plays a vital role in our lives. Earlier, the author has stated that the human brain is hard-wired for logic (which means that logic is a function of the human brain). This implies that if the human brain had been hard-wired differently human logic could have been different too. In logical deduction, the human mind has to strain itself over premises, statements or hypotheses to arrive at the correct connections or conclusions. Earlier too the author has stated that there is the possibility of a mind, perhaps one belonging to an extra-terrestrial (whose brain might be hard-wired differently, if he does possess one), to know without having the need for logical deduction. This would be a super-smart or super-intuitive mind. Our minds are intuitive and to some extent we are capable of sensing and knowing without the need for logical deduction. If only we could have instant understanding and knowledge all the time without having to perform the rigmarole of the act of logical deduction. Our minds would have to be much more intuitive than they are - be super-intuitive. Imagine a mathematician or logician not having the need to use axioms, lemmas, theorems and elaborate logical steps to form valid conclusions. Imagine that through a fantastic leap of intuition the mathematician or logician has instant understanding and knowledge, just like a seer or fortune-teller who could forecast future events before they took place. If our intuition had been much greater, if our brain had been much better constructed, we would perhaps encounter less problems with logic, and, we would probably have created different rules or laws of logic - we would probably have a different kind of logic, perhaps a logic which is much more effective, we might even have a much more hassle-free way of understanding and knowing, e.g., instant understanding and knowledge. In other words, we would have to be much smarter than what we are to be able to reduce or overcome our problems with logic.

How reliable is human logic? Human logic does not seem so reliable after all. The author is of the opinion that the human being should rely more on his physical senses (sight, hearing, touch, taste, smell) than the mere use of logical reasoning. In other words, he should try to find out, to figure out or know, by relying more on his physical senses. Logical deduction might tell us that a certain phenomenon should be the outcome. But, we should verify the outcome with our physical senses, i.e., physically observe whether the outcome would occur indeed if such and such conditions were present, in other words, observe the effect of the cause. Empirical evidence or physical evidence, which is the essence of existence, should rule over logic, which is abstract evidence. Would you prefer to believe something which could be physically observed or something which could not be physically observed? The answer must be obvious. Indeed, seeing is believing! This is precisely the reason why scientists carry out physical experiments (to verify or confirm hypotheses obtained through logical deduction), why manufacturers participate in exhibitions (where their products could be displayed, and, physically studied by potential customers), why judges occasionally visit the scenes of crimes (to physically ascertain for themselves) or have acts of crimes acted out in the court (to physically ascertain certain possibilities), etc. It would indeed be foolhardy to rely entirely on pure logical deduction without carrying out physical verification, especially now that we are now aware of the short-comings of logic.

Therefore, we should always regard logic with an open mind and a healthy skepticism. If only we are so smart that we could understand and know without logical deduction, without cracking our brains. It would not be surprising that such an intelligent race (or races) already exist(s) in another universe (or universes). Our problems with logic hence appear to be the outcome of our faulty mental equipment or brains. For example, due to our imperfect memory, we are easily lost in a long and complex chain of logical reasoning, as by the time we arrive at a later stage of the long chain of reasoning we would normally have forgotten about the earlier stages, resulting in us failing to see the logical link in the long chain of reasoning; by the time we reach the later stages of the long chain of reasoning we would normally have to refer back to the earlier stages in order to see the logical link in the whole chain of reasoning; understanding a long and complex chain of logical reasoning, such as that normally encountered in mathematics, is thus an arduous and time-consuming process requiring a lot of patience; of course a better and more efficient memory would have made the task an easier one; of course, it would have helped in understanding if the long and complex chain of logical reasoning were presented in a clear, simple, precise and easy style. If only human beings were much more intelligent, much smarter. If only our intuitive minds were much more intuitive. If only our memories were much better. Then we would certainly not have so much problems with logic. Human beings should be humble enough to acknowledge the short-comings of their mental faculties. When this happens disputes pertaining to logic would be reduced.

It appears man is generally affected by intellectual pride. He not only thinks he is the smartest animal on earth, many a person thinks that he is smarter than others, many also try to be smart, to outdo others in the smartness game. But, alas, little do these smart alecks realize that logic is imperfect, and, "tricky", that they are just making "much ado about nothing". To these people the author would like to say, "Take it easy. Do not take intelligence and logic so seriously". It has been acknowledged that there are many types of intelligence and that our I.Q. or intelligence tests measure only some of them, which means that a person with a higher I.Q. score is not necessarily more intelligent.

After so much talk and discussion about logic, how important is logic really? If the author might offer his view, it would be, "It does not appear to figure prominently in the behavior of human beings". Human beings could be seen behaving no differently, or, worse, than animals, e.g., butchering each other and blowing each other to pieces. All these are perhaps manifestations of different logics or conflicting logics at work. This implies that human logic is defective and "problematic". Thus, despite our "high level" logic, we are basically brutes, animals.

There appears to be a great barrier to logical thinking or logical behavior, i.e., emotions or passions. It seems that human beings are ruled more by emotions and passions than logic, and, often allow their emotions or passions to color their logic, i.e., the emotions and passions seem to have a greater influence on us than logic. Our logic is often based on prejudiced, biased or faulty premises. The result is confusion, self-contradictions, wrong conclusions, and, of course, problems. When emotions and passions interfere a person might find himself unable to think straight, think clearly. Though it is inherent in man's nature to be emotional and passionate, he should not allow emotions and passions to interfere with his thinking, i.e., he should be as objective as possible, which means that he should be disciplined and should be able to control or rein in his emotions and passions. Many a time a person in a moment of great emotion or passion does something which he later regrets. Therefore, through proper education, a person should have been trained not only in logical thinking, but in self-discipline, in being capable of controlling the emotions and passions. Human beings generally have whims, fancies and urges, and might carry out an action in a fit of fancy or strong emotion, against their better judgement, against logic. So, the emotions and passions seem more important than logic, and being able to deal effectively with them seems very important. Thus, besides learning how to think logically, we should learn how to live balanced, happy and contented lives. What is the point of being intelligent and being able to think logically, if we feel unhappy and discontented all the time? Then, is it not better to be happy and contented though one might not be intelligent or capable of effective logical thinking? Is it not better to be a "happy fool" than an "unhappy Aristotle"? Some might prefer to be "happy fools", while others might prefer to be "unhappy Aristotles" as they value intellect over everything else. The ideal of course is to be smart, happy and contented. It might be argued that if a person is smart he would be able to apply his intelligence to find happiness and contentment, find the objects which bring him happiness and contentment. Yet many apparently intelligent people are clearly unhappy, even depressed. In actuality, a person's happiness has much to do with his environment than anything else. How could a wise and intelligent man surrounded by fools and evil people be really happy? He might attempt to change them for the better with his intellect, but there is no guarantee of success. He might even be unhappy simply because he is smarter band cleverer than his peers, who might find him "queer" and avoid or ostracize him. He might try to make himself more ordinary, less "queer", to be accepted by his more inferior peers - if the acceptance of his peers were so important to him, which is a question of personal taste. It is perhaps only the superiority of a person's intellect which enables him to realize the seriousness of a situation and makes him sad or depressed, while the person of lesser intellect might not even be aware of the seriousness of the situation and hence would be able to sleep over it - remember the adage: Ignorance is bliss.

But, intelligence and logic are important for the success of any human enterprise and the survival and progress of the human race. And, intelligence, aka logic, could be used for good as well as for evil. Let us hope that there would never be any evil geniuses out there to wreak terror and havoc in society. Let us hope that society would be capable enough to prevent such an untoward situation from arising and good enough to see to it that intelligence would always be used for good and peaceful purposes. It would boil down to the question of good triumphing over evil, intellectually, and, physically.

Cheers to logic for the good and betterment of mankind, a logic which is noble, pure, objective and indisputable. Nay to twisted, crooked or pseudo logic.

A few blunt remarks on logic: What is the damned use of logic if, instead of helping us solve our most pressing problems, especially problems pertaining to inter-human relationships, it helps to create rifts in our society, with the result that there are conflicts galore? This would probably be logic at its most counter-productive and lowest ebb. It might then be better off to be without logic. And, logic might not even be a pre-condition or pre-requisite for survival. For instance, a tree could survive for hundreds of years or more without any apparent need for logic, e.g., through adaptation to its environment. (Being able to adapt to one's environment is a sign of intelligence in a human being. So, could a tree not be regarded as intelligent since it could adapt to its environment?) Could man with his logic ever hope to emulate this? Is it better to be not too reliant on logic? Is it not better to have a free and open mind on logic and regard it with a healthy skepticism? At this point here, we should all know that logic is imperfect and is not to be taken too seriously. To be ignorant of this is to be so at our own peril. Perhaps, we would be really better off if all our brains had been hard-wired to see logic in only one way (instead of so many different ways, as it is apparently so now), just as our physical senses make us all, e.g., see the color green as green or red as red. (Imagine the chaos if our physical senses make us see colors differently, e.g., red might be red to some, blue to some others, green to yet others, and so on. How ridiculous it would then be! How disruptive!) Enough is said. The lesson should have been learnt.

In lieu of the human brains being hard-wired to see logic in only one way, to think alike, they could be coerced to do so. Our society functions with hierarchy, a social structure. At the very top, we have the leaders and governments, who lay down the rules or laws which have to be adhered to by everyone down the social ladder. (Hierarchy and rules or laws ensure some form of conformity and uniformity, and, preserve some necessary order and prevent chaos.) These leaders and governments could of course lay down the rules or laws pertaining to all matters related to logic or logical reasoning so that disputation does not exist or is minimal. Alternatively, we could have a carefully and democratically elected intellectual elite who are, theoretically, trained and experts in logic and the art of logical deduction to draw up codes of conduct and rules pertaining to the use of logic, which are to be adhered to by everyone. In other words, we should attempt to get everyone to "think alike". Does this not mean that originality would be bad? Originality, if it "rocks the boat" and causes disputes and problems would of course be undesirable. All this might be easier said than done. But, to rid society of the various schisms and rifts which threaten to break it up something to the effect described here should be attempted. As they say, if there is no venture there would not be any gain. If there are no laws to guide and control the actions of people, everyone would do as they like without any concern for the general well-being of everyone in the group, in society, and, the result would be utter chaos, pandemonium. In short, if people were allowed to let their thoughts, their logic, run wild, there would be terrific chaos, terrible disharmony.

The author would like to say something more about the above-mentioned carefully and democratically elected intellectual elite. This specially elected group of intellectuals should be capable of producing the logic tables and logic guiders, a kind of regulations or laws for everyone to follow, as well as ensuring that they are effectively implemented. As society needs laws and regulations in order to function in an orderly manner, it also needs laws and rules for regulating intellectual activities or the activities of brain-workers. With rules and guidelines to abide by or refer to there should be no necessity to engage in disputes or arguments. Arbitration of disputes and court cases might now be the norm for unsettled disputes. But, with the effective implementation of these logic tables and logic guides, arbitration and court hearings should more or less be a thing of the past. These logic rules and regulations should have the objective of being as fair as possible to all the parties concerned as well as being able to bring as much good as possible and cause as little harm as possible to the society at large. In practically all other aspects of human activities there are rules and regulations to act as guide, e.g., we have rules and regulations in the game of chess, soccer, badminton and golf. But, why are rules and regulations lacking in the more serious activities such as intellectual activities, or activities involving the use of logic? This appears to be some sort of discrepancy. We might have the various systems of logic created by the logicians and mathematicians, but, unfortunately, these systems appear arbitrary and difficult to put into practice though on paper they might appear fine. Hence, the logic tables and logic guides should be convincing, acceptable and easy to put into practice. They should be legislated in order to have an effective, beneficial impact on the society at large. They should have an orderly effect on the way intellectual activities are carried out. Original ideas, or even revolutionary ones, should still be welcomed, but should not be allowed to be forced upon others; they should only be adopted by others when the majority approves them and wishes to have them adopted. So, in the intellectual sphere, in the realm of logic, the principles of democracy could still be put into practice. And, democracy assumes that people are matured and wise enough to know what is best for themselves, and, hence, are able to elect or choose the best or most appropriate options for themselves. But, this need not be the case. What happens if the people who are to choose, elect or vote are really lacking in wisdom and maturity? In that case, we might have to be sorry for society, for the way activities, including intellectual activities and activities involving logic, would be carried out. There is a theory which states that the group is generally wiser and more capable of making better decisions than the individual. Let us hope that this theory would hold true, or, there could be problems in a democracy, when people are not wise or matured enough to exercise their voting rights in the most appropriate manner. What happens if they elect the wrong leaders to run society? It seems dicey. Perhaps, the people who elect the leaders should be qualified first, i.e., meet certain standards or criteria, e.g., educational level, intelligence, moral character, etc. The people who elect the leaders who run society should not be just any Tom, Dick or Harry. They should either be carefully selected or go through a proper election process. If everyone, including morons, immoral people and criminals, could just cast their votes and elect their leaders, might it not happen that inappropriate or wrong leaders could end up running the show and causing more problems than solving them? This appears to be a chink in the armor of democracy. Let us stop here now and not digress further.

Logic has also been beset by paradoxes, which indeed show up its weakness. As mentioned earlier, logical reasoning should be carried out with true statements if the conclusion or result were to be true. Deduction based on untrue statements would of course end up with a conclusion which is also untrue. Unfortunately, it is not always possible to differentiate truths from untruths, and, many a time, the premises for a logical deduction have to be assumed to be true, if it were not possible to ascertain their trueness, if the process of logical deduction were to proceed at all. A typical example is mathematical reasoning, where axioms (obvious, assumed, but unproven, true statements), and, lemmas and theorems (which are statements that have been proven to be true) are used as premises.

The author is of the view that paradoxes are the result of reasoning with untrue premises or premises not based on reality or real facts. Take for example, the statement, "This statement is false". This statement is self-contradictory. If this statement is true, it is not false, and, if it is false, it is true. In reality, in real life, would anyone ever pass such a remark, except in jest? There is another statement, which has tickled the author. It is, "I am humble and am proud of my humility". In this case, am I proud or humble? I seem to be contradicting myself when I make this remark - on the one hand, I say that I am "humble", then I contradict myself by saying that I am "proud" (of my humility). There is yet another interpretation: I am humble about other things beside my humility, but I am proud of my humility (only). Consider the following statement:

Man is so intelligent that he could make his stupidity look like intelligence.

This statement is self-contradictory. So, is man really intelligent, or, stupid? Thus, logic has a great deal to do with the meanings of words and their interpretation, which implies that if logic were to be problem-free, or paradox-free, the words, terms and premises used in the process of logical deduction should be carefully and precisely chosen - for precision of thought, the correct, proper words or terms are a necessity - that is, the words or terms used should be precise in meaning, and not have subtle or multiple interpretations. The author would like to emphasize again that the mastery of a proper language, or, a proper set of symbols (especially for mathematicians), is necessary for effective logical reasoning.

Another aspect of logic involves generalisation, or, forming conclusions. For example, mathematicians are known to generalise things with their mathematical logic. Unfortunately, in the real world, things are not that simplistic, things are much more complex than that. A mathematician might create a mathematical model of how the economy behaves; there is this branch of mathematics known as econometrics, which uses mathematics to carry out forecasts of the economy. But, such a model is at best only a simplistic model, which might have little or no relevance to how the economy would actually behave. In reality, the economy is probably much too complex for an accurate forecast to be possible. In any case, to have a forecast, a plan, even an unlikely to be accurate one, is better than to have none at all. This is one of the shortcomings of logic. It should be stated that though logic works on the basis of "cause and effect", it appears to fail miserably where there is chaos. Chaos is the situation where a slight perturbation, interference, causes an "out-of-portion" or overwhelming result or effect, and, chaos theory is that relatively new branch of mathematics where mathematicians are trying to come to terms with the fact that in chaos it is practically impossible to make any forecasts or predictions at all. The thoughts and ideas of human beings appear at times to be characterised by chaos, and, that is when logic would not be of use.

Hence, logic would seem to work only when ideas and thoughts conform to certain accepted rules or conditions, certain conventional wisdom. Thus, a person whose ideas and thoughts are too original and do not conform to conventional wisdom is likely to find himself not understood, and, worse, he might be misunderstood. It appears to be a case like "we are logical and right in our ideas and thoughts, and, therefore, anyone whose ideas and thoughts are different are wrong and not logical". Unfortunately, logic is not as simple as this, and, what is logical or sensible to one sane person might not be so to another sane person, i.e., logic is arbitrary. And, how do we judge or prove sanity or soundness of mind? This is yet another hurdle. As logic is a function of the sanity or soundness of mind of the person, does it mean that there are at least two kinds of logic: the logic of the person of sane mind and the logic of the person of insane mind? Here, we probably have a situation wherein to the person of sane mind the person of insane mind is illogical and to the person of insane mind the person of sane mind is illogical. If a third party of sane mind were to arbitrate a logical dispute between the two, what would be the result? On the other hand, if a third party of insane mind were to arbitrate a logical dispute between the two, what would be the outcome? Indeed, it could be a situation wherein to the sane person the insane person is insane, and, to the insane person, the sane person is insane. What happens if the psychiatrist were of insane mind? Would he treat a person of sane mind as a person suffering from insanity? Indeed, how do we prove sanity, or, insanity? How do we distinguish between the logic of the sane and the logic of the insane; what are the differences, if any, between the two? Perhaps, it is all relative, logic is relative.

Logical reasoning is one of the ways for us to have knowledge. Other ways of knowing are, of course, by using our physical senses, e.g., seeing and hearing. We could see, hear, or witness, something, and we would then know that it is true. However, if there were no opportunity to see, hear or witness an incident, we could still arrive at the conclusion that the incident had occurred by the process of logical deduction or logical reasoning. But, then, we normally still have to "physically confirm" our conclusion which has been arrived at by the process of logical reasoning (as in the case of the scientist at work). That is why the author has earlier stated that one should rely more on one's physical senses than on mere logic. In other words, it is better and easier to deal with the more concrete things of the real, actual world than the abstract entities of logic.

When we consider the logic of ideas or thoughts we are in effect thinking of what these ideas or thoughts imply, what other ideas or thoughts these ideas or thoughts are linked to or connected with. For example, we think of A. Then we realize that A implies B or is linked to B, B implies C or is linked to C, and so on and so forth. The result here would be a chain of reasoning. To be familiar with effective and result-oriented reasoning or deduction, the reader should peruse the Sherlock Holmes stories created by Arthur Conan Doyle. The author himself has read practically all these stories and has thereby definitely gained much insight of logical deduction. Though Sherlock Holmes is a fictitious character, his creator, Arthur Conan Doyle, who was himself a medical doctor, had modeled him after a real-life medical doctor, who had acute powers of observation and deduction, by the name of Joseph Bell, a person that Doyle had admired.

The author has indeed been interested in many subjects where logic plays an important role, e.g., mathematics and its so called mathematical logic, which could appear deep, profound, complex, abstruse and incomprehensible but is generally impractical, the physical sciences such as physics where the logic appears to be of a more practical or applied nature (in fact, there have been instances of mathematicians who found pure mathematics (as opposed to applied mathematics) "useless" and having little practical applications if any at all switching over to the sciences such as physics where their mathematical knowledge could be put to good, practical use), the social sciences such as economics and psychology where the logic has a more practical, mundane and "human interest" slant, and, philosophy (the author's favorite) which is concerned with the truth of everything, which is to be arrived at through analysis and the use of logic. The author has been especially keen on mathematics, whose symbolism is often arcane and impenetrable, but which has a claim to rigor or a high standard of logical precision. Being able to solve a very difficult mathematical problem might give one the "high", like being able to play excellent chess and being able to beat a chess grandmaster. Mathematics and chess might provide some good training for thinking and bring some self-satisfaction and pleasure, but they could not directly benefit society unless they could be "applied" to solve or tackle our practical problems, our real problems, e.g., in the case of applied mathematics such as statistics and operations research. For example, consider the author's solutions to the two very difficult problems in pure mathematics, viz., the twin primes conjecture (Is there an infinitude of twin primes?) and the Goldbach conjecture (Is there an infinitude of even numbers which are each the sum of two prime numbers?). The author has in his solutions proved that there is an infinitude of twin primes and an infinitude of even numbers which are each the sum of two primes. The question is, "What does it matter at all?" The author might be happy for a while that he has the solutions, one of which has been published in a respected mathematics journal. Some others who had perused his solutions might appreciate them and be pleased for a while. These solutions might have a little "entertainment value". Of course, without them, we could all certainly find other things to please us, things which are even likely to please us more, things which have higher "entertainment value". Without the prime numbers, without the even numbers, even without mathematics, life would still be able to go on. There are certainly many people with practically zero mathematical knowledge who have lived successful, happy lives, lives which are much more successful and happier than those of the top mathematical geniuses. On the contrary, many mathematical geniuses and famous ones too, e.g., Galois, Cantor and Riemann, had lived tragic lives. The author is here not crying "sour grapes" because he is not a mathematical genius. He is just stating a plain, simple fact.

Logic per se is in fact a waste, of no importance, if it could not be applied to the affairs of the society at large, if it could not be put to practical use. It would then just be another beautiful "work of art", which is more or less on the par with a beautiful piece of painting or sculpture, or a beautiful musical piece. On the other hand, it seems that logic has often been wrongly applied, i.e., it has often been twisted, manipulated, exaggerated and forced upon others, resulting in problems galore (whereby one man's logic could be another man's nonsense). So far, in this tome, the author has been painting the picture of logic as a troublesome, imperfect creature, which needs to be handled with care, an open mind and a healthy skepticism. Here, the author is suggesting that we should be more concerned with applied logic than pure logic, i.e., logic should be result-oriented.

Concluding here, we could say that logic has an important role to play in our search for the truth. Of course, the truth could be directly witnessed through the physical senses (sight, hearing, touch, taste and smell, especially seeing, whereby seeing is believing). Where the truth could not be directly found through the physical senses, logical deduction might be able to help us to arrive at the truth. Thus, the predominance of the physical senses vis-à-vis the faculty of logical reasoning.

3

APPLICATION OF NAVIER-STOKES EQUATIONS

The Navier-Stokes differential equations describe the motion of fluids which are incompressible. The three-dimensional Navier-Stokes equations misbehave very badly although they are relatively simple-looking. The solutions could wind up being extremely unstable even with nice, smooth, reasonably harmless initial conditions. A mathematical understanding of the outrageous behaviour of these equations would dramatically alter the field of fluid mechanics. This chapter describes why the three-dimensional Navier-Stokes equations are not solvable, i.e., the equations cannot be used to model turbulence, which is a three-dimensional phenomenon.

The general equations of motion for a viscous fluid were obtained by Sir George Stokes in 1845. The following is the fundamental equation (in vectorial form) governing the flow of a viscous fluid:-

$$\frac{\partial v}{\partial t} + (v \cdot \nabla)v = -\frac{1}{\rho}\nabla Pe - \nabla \varphi + \frac{\eta}{\rho}\nabla^2 v,$$

where v is the velocity of the fluid (as a function of position), Pe the pressure, φ the gravitational potential, ρ the density and η the viscosity.

A fluid in motion could be characterised by its velocity field (velocity as a function of position). However, because of the complex nature of the forces affecting fluids (in general, forces of both compression and viscosity) the result of applying basic principles such as Newton's second law is a set of nonlinear equations. Computational methods therefore play a large part in fluid dynamics. (Newton's second law states that the rate of change of momentum p of a body equals the total force F acting upon it, as is described by the following equation:-

$$F = \partial p / \partial t$$

If, as is normally the case, the mass of the body is constant, $F = \partial(mv)/\partial t$ reduces to $F = m\partial v/\partial t$ or $F = ma$, where a is the acceleration of the body. Note that the force and acceleration are vectors. The first law is the null case of the second law (if $F = 0$ then $a = 0$).)

The Navier-Stokes equation is a miracle of brevity, relating a fluid's velocity, pressure, density and viscosity. In two dimensions, fluid flow governed by this partial differential equation is deterministic and predictable. But this equation fails when the fluid becomes turbulent as turbulence represents three-dimensional flow of the fluid, for which the Navier-Stokes equation does very poorly. Whereas fluid flow under normal conditions tends to be laminar, in turbulence it becomes irregular and develops eddies, ripples and whorls. But yet there is some sort of order found within this disorder or turbulence which could be described as self-similar or fractal. What mathematical technique could be used to describe this state?

The Navier-Stokes equations are nonlinear and do not submit to any general method of solution. Each new problem has to be carefully formulated as to geometry and proper boundary conditions. Then some scheme of attack might be adopted with the hope of reaching a solution. In most cases all attempts to obtain an exact solution fail. Approximate solutions have to make do. In a few cases exact solutions could be obtained. The possibility that perhaps the flow of the fluid is unidirectional, i.e., v (x, y, t) = 0, is not an assumption. It is rather an intuitive guess which is pursued until we either find a solution or become convinced that it does not lead to a solution, in which case we mark it as an unsuccessful trial.

Substitution of viscosity in the Navier-Stokes equations with viscosity = 0 reduces them to a form called the Euler equations:-

$$p \frac{Dq}{Dt} = pg - \nabla p \quad \text{(in vectorial form)}$$

The Euler equations had been formulated earlier than the Navier-Stokes equations and considered an approximation. The Euler equations are of the first order and cannot in general satisfy the boundary conditions. We could therefore conclude that the Euler equations do not form a good approximation near a rigid boundary. Far from a boundary and where viscosity = 0 is a fair estimate, they have an important role as approximations and are generally easier to solve than the full Navier-Stokes equations.

The Navier-Stokes equations do need for their solution initial conditions as well as boundary conditions. The following are proper boundary conditions for a velocity on a rigid boundary:-

$$q_n = q_t = 0 ,$$

where q_n is the normal component of the velocity relative to the solid boundary, and q_t is the tangential component. These conditions are also termed the no-penetration ($q_n = 0$) and no-slip ($q_t = 0$) viscous boundary conditions. When the region occupied by the fluid is not closed, i.e., the fluid is not completely confined, additional conditions are still required on some surfaces which completely enclose the domain of the solution. These might represent some real physical surfaces or they might be chosen quite arbitrarily, provided the velocity on them is known. The pressure, which is also a dependent variable, also requires boundary conditions. The Navier-Stokes equations are then satisfied and we now know the resulting pressure field. This flow can exist only if the obtained pressure is possible. An acceptable boundary condition might be: $p = p_\infty$ = const at $r \quad \infty$, which then implies: $p = p_\infty - pQ^2/8\Pi^2 \cdot 1/r^2$. We also note that in the for the pressure there is no trace of the viscosity. This pressure therefore also satisfies the Euler equations. (As viscosity in a fluid enables it to smooth out or overcome the ripples, eddies and whorls of turbulence, a viscous fluid is in effect not so much affected by turbulence than a non-viscous fluid. Thus, the Navier-Stokes equations, as they relate to viscous fluids, present a better solution for incompressible fluids which are viscous and subject to turbulence than the Euler equations for non-viscous fluids.)

The scientist normally makes a forecast of the outcome of a flow and uses the Navier-Stokes equations to model this forecast. However, in the instance of turbulence, making this forecast will be fraught with difficulty, if it can be carried out at all. Putting it another way, if turbulence could be forecasted, predicted and described by the Navier-Stokes equations it could not be turbulence, for turbulence implies puzzlement, lack of order or pattern and lack of predictability.

The Navier-Stokes equations are nonlinear due to the acceleration terms such as $u\partial u/\partial x$. As a result, the solution to these equations may not be unique. For instance, the flow between two rotating cylinders can be solved using the Navier-Stokes equations to treat a relatively simple flow with circular streamlines; it can also be a flow with streamlines which are like a spring wound around the cylinders as a torus; there are also more complex flows which are solutions to the Navier-Stokes equations, all satisfying the identical boundary conditions.

For simple geometries, the Navier-Stokes equations can be solved with relative ease. However, the equations cannot be solved for a turbulent flow even for the simplest of examples. A turbulent flow is highly unsteady, nonlinear and three-dimensional and therefore requires that the three velocity components be specified at all points in a region of interest at some initial time, say $t = 0$. But, even for the simplest geometry, such information will be almost impossible to obtain.

Therefore, the solutions for turbulent flows have to be left to the experimentalist and are not attempted by solving the Navier-Stokes equations.

4

NAVIER-STOKES EQUATIONS AND LIMITATION

The motion of fluids which are incompressible could be described by the Navier-Stokes differential equations. However, the three-dimensional Navier-Stokes equations for modelling turbulence misbehave very badly although they are relatively simple-looking. The solutions could wind up being extremely unstable even with nice, smooth, reasonably harmless initial conditions. A mathematical understanding of the outrageous behaviour of these equations would greatly affect the field of fluid mechanics. A reasoned, practical approach towards resolving the issue and a practical, statistical kind of mathematical solution are brought up.

Navier-Stokes Equation

The general equations of motion for a viscous fluid had been obtained by Sir George Stokes in 1845. The fundamental equation (in vectorial form) governing the flow of a viscous fluid is as follows:-

$$\frac{\partial v}{\partial t} + (v \cdot \nabla)v = -\frac{1}{\rho}\nabla Pe - \nabla \varphi + \frac{\eta}{\rho}\nabla^2 v,$$

where v is the velocity of the fluid (as a function of position), Pe the pressure, φ the gravitational potential, ρ the density and η the viscosity.

A fluid in motion could be characterised by its velocity field (velocity as a function of position). However, because of the complex nature of the forces affecting fluids (in general, forces of both compression and viscosity) the result of applying basic principles such as Newton's second law is a set of nonlinear equations. Computational methods therefore play a large part in fluid dynamics. (Newton's second law states that the rate of change of momentum p of a body equals the total force F acting upon it, as is described by the following equation:

$$F = \partial p/\partial t.$$

If, as is normally the case, the mass of the body is constant, $F = \partial(mv)/\partial t$ reduces to $F = m\partial v/\partial t$ or $F = ma$, where a is the acceleration of the body. Note that the force and acceleration are vectors. The first law is the null case of the second law (if $F = 0$ then $a = 0$).)

The Navier-Stokes equation is a miracle of brevity, relating a fluid's velocity, pressure, density and viscosity. In two dimensions, fluid flow governed by this partial differential equation is deterministic and predictable. But this equation fails when the fluid becomes turbulent as turbulence represents three-dimensional flow of the fluid, for which the Navier-Stokes equation does very poorly. Whereas fluid flow under normal conditions tends to be laminar, in turbulence it becomes irregular and develops eddies, ripples and whorls. But yet there is some sort of order found within this disorder or turbulence which could be described as self-similar or fractal. What mathematical technique could be used to describe this state?

The Navier-Stokes equations are nonlinear and do not submit to any general method of solution. Each new problem has to be carefully formulated as to geometry and proper boundary conditions. Then some scheme of attack might be adopted with the hope of reaching a solution. In most cases all attempts to obtain an exact solution fail. Approximate solutions have to make do. In a few cases exact solutions could be obtained. The possibility that perhaps the flow of the fluid is unidirectional, i.e., v (x, y, t) = 0, is not an assumption. It is rather an intuitive guess which is pursued until we either find a solution or become convinced that it does not lead to a solution, in which case we mark it as an unsuccessful trial.

Substitution of viscosity in the Navier-Stokes equations with viscosity = 0 reduces them to a form called the Euler equations:

$$p \frac{Dq}{Dt} = pg - \nabla p \quad \text{(in vectorial form)}$$

The Euler equations had been formulated earlier than the Navier-Stokes equations and considered an approximation. The Euler equations are of the first order and cannot in general satisfy the boundary conditions. We could therefore conclude that the Euler equations do not form a good approximation near a rigid boundary. Far from a boundary and where viscosity $= 0$ is a fair estimate, they have an important role as approximations and are generally easier to solve than the full Navier-Stokes equations.

The Navier-Stokes equations do need for their solution initial conditions as well as boundary conditions. The following are proper boundary conditions for a velocity on a rigid boundary:

$$q_n = q_t = 0 \; ,$$

where q_n is the normal component of the velocity relative to the solid boundary, and q_t is the tangential component. These conditions are also termed the no-penetration ($q_n = 0$) and no-slip ($q_t = 0$) viscous boundary conditions. When the region occupied by the fluid is not closed, i.e., the fluid is not completely confined, additional conditions are still required on some surfaces which completely enclose the domain of the solution. These might represent some real physical surfaces or they might be chosen quite arbitrarily, provided the velocity on them is known. The pressure, which is also a dependent variable, also requires boundary conditions. The Navier-Stokes equations are then satisfied and we now know the resulting pressure field. This flow can exist only if the obtained pressure is possible. An acceptable boundary condition might be: $p = p_\infty = $ const at $r \quad \infty$, which then implies: $p = p_\infty - pQ^2/8\Pi^2 \cdot 1/r^2$. We also note that in the solution for the pressure there is no trace of the viscosity. This pressure therefore also satisfies the Euler equations. (As viscosity in a fluid enables it to smooth out or overcome the ripples, eddies and whorls of turbulence, a viscous fluid is in effect not so much affected by turbulence than a non-viscous fluid. Thus, the Navier-Stokes equations, as they relate to viscous fluids, present a better solution for incompressible fluids which are viscous and subject to turbulence than the Euler equations for non-viscous fluids.)

Newton's Law of Viscosity

The following equation is known as Newton's law of viscosity:-

$$T_{yx} = \mu \frac{du}{dy}$$

A fluid that obeys this law is called a Newtonian fluid. This equation states that in unidirectional flow the shear stress in a Newtonian fluid is directly proportional to the transverse velocity gradient, du/dy, which is also known as the rate of shear strain or the rate of shear deformation.

There is no obvious reason why real fluids should obey this law. In fact, there are more fluids that do not obey this equation than those that do. Fluids that do not obey this law are called non-Newtonian. It is fortunate that the three most abundant fluids, air, water and petroleum, obey Newton's law of viscosity rather closely. The typical non-Newtonian fluids are paints, polymer solutions and melts, blood and many liquid food products, such as jellies, soups, etc.

Computing of Reynolds Numbers

The following is the formula for the Reynolds number:-

$$Re = vLp/n,$$

where p (kg/m^3) is the fluid's density, v (m/s) is a typical fluid velocity, L (m) is some characteristic length, e.g., diameter of a pipe through which the fluid flows, and, n (kgm/s^2) is the coefficient of viscosity (Gooey substances have a higher n value than runny ones like petrol. Viscous force = ability to smooth out turbulent whorls, eddies and ripples.).

The Reynolds number is the ratio between two forces, the initial force and the viscous force (frictional force). When the Reynolds number is below a few hundred, the flow of the fluid is smooth. When the Reynolds number exceeds about 2000 to 3000, the flow is completely turbulent. Between these values, the flow is sometimes smooth, sometimes turbulent. The greater the viscosity of the fluid is, the greater is its capability in overcoming or smoothing out the whorls, eddies and ripples of turbulence.

Fluid Flow

According to the Law of Continuity, water flowing from a wide pipe to a narrower pipe speeds up, while water flowing from a narrow pipe to a wider one slows down, and, the slow-moving water in the wide pipe would always have a higher pressure than the fast-moving water in the narrower pipe. Turbulent fluid flow could cause a pipe to give way.

Some equations pertaining to fluid flow are as follows:-

Energy (E) = Mass (m) x Speed (v^2) (i)

Altitude (A) + Energy (E) = Constant (ii)

Energy (E) = Density of Fluid (p) x Speed (v^2) (iii)

Pressure (P) + Energy (E) = Constant (iv)

Pressure (P) + Density of Fluid (p) x Speed (v^2) = Constant (v)

Model for Turbulence

The logistic equation has been a popular method used for the modelling of turbulence and chaos. The formula had been invented by the Belgian mathematician, Pierre Francois Verhulst, and is very simple. But because the process has to be repeated over and over again it ends up being extremely complicated. The equation works in the following way:-

For instance, if X is the population now, then the population next year is given by:

$$X_{next} = rX(1-X),$$

where r is some constant which could be adjusted according to the population being modelled. It is simplest if values of X between 0 and 1 are taken, so that 1 is the maximum population and 0 represents extinction. We might, for instance, take an arbitrary value for r of 2.6, and begin thus.

Let $X = 0.2$. Then $1 - X = 0.8$, and, $X(1 - X) = 0.2 \times 0.8 = 0.16$.

Then multiply this result by 2.6 and we would get 0.416.

Repeat the process. Start with $X = 0.416$ and we would get 0.6317. The population increases.

Start with 0.6317 and we would get 0.6049. The population falls.

Start with 0.6049 and we would get 0.6214. The population goes up again.

Repeating or iterating this process over and over again we would obtain the following population figures:

0.6117, 0.6176, 0.6141, 0.6162, 0.6150, 0.6156, 0.6152, 0.6155, 0.6153, 0.6154, 0.6153, 0.6154, 0.6154, 0.6154.

The population rises and falls but converges on a fixed number.

Scientists have however tried to model turbulence or chaos using this equation. But, does this model really describe turbulence or chaos?

Analysing Motions of Fluids with Fourier Analysis

The surface water in a wave moves in a circular path at an angular velocity $w = 2\pi/T$ where T is the period of rotation. Deeper water moves in ellipses of decreasing size and increasing eccentricity.

The superpositions of simple sinusoidal oscillations in fluids could produce more complicated patterns of oscillations. The inverse mathematical operation of Fourier analysis could reduce any complicated oscillation into a sum of its simple sinusoidal components, each with a different period and amplitude.

With waves, there is a phenomenon which oscillates both in time and space. It might seem that this would considerably complicate any mathematical attempt to describe the superposition of waves. We could practically analyse complicated wave shapes either by freezing them in time or by freezing them in space. In the time domain, we obtain the wave's frequency components, while in the space domain we get the corresponding spectrum of wavelengths. These two approaches could each stand on their own, one being transformable to the other, because the product $f\lambda$ is a constant (the wave speed), the longer wavelengths corresponding to the lower frequencies (i.e., the longer periods). The wavelength spectrum could be computed directly from the frequency spectrum by noting that for each harmonic component $\lambda = v/f = v/T$ - this holds for all of the wave's Fourier components (e.g., $\lambda_o = 3$ metres, $f_0 = 20$ cycles per second and wave speed $v = 60$ m/s).

Fourier Series and Circles

We could interpret a Fourier series geometrically as the projection of a system of superimposed circular motions. A circle rides upon the nest of spinning circles beneath it, and, we project the motion of a point on its circumference onto a line. The result is a periodic but distinctly non-sinusoidal motion which could be described mathematically as follows:-

$$y(t) = \frac{(1)}{T}\sin\frac{(2\pi t)}{T} + \frac{(1)}{3}\sin\frac{(3.2\pi t)}{T} + \frac{(1)}{5}\sin\frac{(5.2\pi t)}{T} + \frac{(1)}{7}\sin\frac{(7.2\pi t)}{T}$$

where T = fundamental period of oscillation = time for one rotation of the largest circle

This is the Fourier series which has been truncated after the fourth term for greater simplicity. If more terms are added (i.e., more circles turning upon circles), the resulting graph would more nearly approximate a series of alternating horizontal line segments.

Probability Waves and Turbulence

We now examine several related important ideas in quantum theory. Schrodinger had found an equation that could be applied to any physical system in which the mathematical form of the energy is known, which is as follows:-

$$\frac{\partial^2 \Psi}{\partial x^2} + \frac{8\pi^2 m}{h^2}(E - V)\psi = 0$$

where ∂^2 is the second derivative with respect to x, x is the position of the particle, ψ is the Schrodinger wave function, or, the probability amplitude for an electron in the state n to scatter into the direction m, E is energy and V is potential energy.

The Schrodinger equation is a deterministic time-symmetrical description of nature. In classical mechanics, when one says that a quantum system is in a particular "state", one means that the state is a point in phase space. It is here described by a wave function whose evolution over time is expressed by the following equation:-

$$ih/2\pi \, \partial\psi(t)/\partial t = H_{op}\psi(t)$$

This equation identifies the time derivative of the Schrodinger wave function ψ with the action of the Hamiltonian operator on ψ. It is not derived but assumed at the start, and could thus be validated only by experiment. In quantum theory, it is the fundamental law of nature. Here, ψ is the probability amplitude for an electron - it is only an abstraction and has no physical reality. ψ is also, in a sense, the electron's own intensity wave. When it is squared and the absolute value is taken, it turns out to be a physical probability of the associated particle's presence.

Later, Born stated that the probability of the existence of a state is given by the square of the normalised amplitude of the individual wave function (i.e., ψ^2). This was another new concept, i.e., the probability that a certain quantum state exists. Born had said there were no more exact answers in atomic theory, but just probabilities. The wave Ψ determines the likelihood that the electron would be in a particular position, and has no physical reality unlike the electromagnetic field.

Dirac had posited that light could be treated as waves or particles. In fact, in quantum mechanics, particles are regarded as waves. The behaviour of these particles could be predicted, as it were, and, they are therefore known as probability waves or Dirac wave particles. A wave/particle duality is present. However, when the particle is not observed, it remains a wave (a probability wave), but upon being observed it becomes a particle.

The following is the formal solution of the Schrodinger equation:-

$$\psi(t) = U(t)\psi(0)$$

where $U(t) = e^{iHt}$, U(t) is the evolution operator which links the value of the wave function at time t to that at the initial time $t = 0$. Both future and past play the same part, since $U(t_1) U(t_2) = U(t_1 + t_2)$, whatever the sign of t_1 and t_2. This property defines a dynamical group.

This description of how quantum particles behave could not be strictly applied to the macro world of fluid flow. Despite this, the above "probability" principles pertaining to how quantum particles behave could be somewhat broadly adapted and used as a guide for the interpretation of turbulent fluid flow.

In another important theory, the Uncertainty Principle, which was propounded by Heisenberg, it is posited that the very act of observing a quantum particle affects its behaviour. According to this theory, the position and the momentum of an elementary particle could not be known simultaneously. The reason for this is that if an electron could be held still long enough for its position to be determined, then its momentum could no longer be determined. A special point is that the product of two uncertainties (or spreads of possible values) is always at least a certain minimum number. From the de Broglie/Einstein relation, $\Delta p \sim h/\lambda$, Heisenberg obtained the imprecision in the momentum. Multiplying the two inaccuracies together, he showed that the product, $\Delta x \Delta p$, would always be greater than or equal to ($\geq$) a certain amount, which is as follows:-

$$(\Delta x)(\Delta p) \geq (\lambda)(h/\lambda) \geq h, \text{ or, } \ldots \quad (i)$$

$$\Delta x \Delta p \geq h \quad (ii)$$

where Δp and h/λ represent the de Broglie relation, and, Δx and λ are from the diffraction limit.

The frustrated researcher looking for certainty must always make a compromise, knowledge gained about time, for instance, is paid for in uncertainty about frequency and vice versa. Though one does not notice Heisenberg Uncertainty Principle in one's everyday experience with the gross macroscopic world, the wave/particle duality defeats the atomic experimentalist who looks for perfection.

Another important implication of uncertainty which is worthy of comment is its effect on causality - the relation of cause to effect. Cause produces an effect. In classical physics if one understands fully the nature of a particular cause, one could then predict the effect. Cause and effect and predictability were cornerstones of classical physics and now they were under question. If it is impossible to measure precisely both the position and velocity of an electron (or any other particle) at the same moment, then it is also impossible to predict exactly where that electron would be at any given time afterward. An experimenter could send off two electrons in the same direction, at the same speed, and they would not necessarily end up in the same location. In the language of physics, the same cause could bring about different effects. There are serious philosophical consequences in this idea.

All this in effect represents "chaotic" behaviour in the quantum world. We have to remember that chaotic behaviour is beyond prediction; the above-mentioned describes the behaviour of quantum particles in a probabilistic, uncertain sort of way - it represents a branch of physics known as statistical mechanics. The corollary of this is the macro-world phenomenon of three-dimensional turbulent fluid flow which defies the solution of the Navier-Stokes equations. Perhaps, a probability function, Φ, should be incorporated in the Navier-Stokes equation, like the case for the quantum particles.

Nature of Turbulence

It has been found that the transition from steady state through several splittings or bifurcations to chaos is similar to phase transitions - transitions that take place when a substance changes from a gas to a liquid or a liquid to a solid - all these transitions are also similar in that scaling is involved. It is believed that the presence of strange attractors, which could be defined as an endless path in phase space where the future depends sensitively on the initial conditions, is responsible for the presence of turbulence or chaos. A strange attractor possesses the following characteristics:-

i) It is generated by a simple set of differential equations.
ii) It attracts and therefore all nearby trajectories in phase space converge toward it.
iii) It has a great or very sensitive dependence on initial conditions, i.e., tiny differences or errors in the initial conditions lead quickly to large differences in the trajectory.
iv) It is fractal, i.e., there is self-similarity or some familiar pattern within it.

The problem with fluid flow in conditions of turbulence is that the path taken by the fluid is continuous but nowhere differentiable. Turbulence gives rise to whorls, eddies and ripples in the fluid. However, there is a self-similar structure or pattern within the fluid - whorls would be found within whorls. This is in accordance with the well-established self-similarity concept which had been developed by Mitchell Feigenbaum in the 1970s and which brought him fame. According to this concept, there is a tendency of identical mathematical structures to recur on many levels, and, within a given structure there would be smaller copies of the same structure, their sizes being determined by the scaling factor, which is 4.669 and found to be a constant like pi (3.142). (This means there is some kind of order or pattern found in turbulence, chaos or disorder.) If we were to plot a curve to describe the fluid's movement under conditions of turbulence we could expect the curve to be rough and nonlinear (which means it is not possible to derive the differential equations for describing this curve, thus making predictions concerning the fluid's movement very difficult if not impossible). However, viscous fluids, for which the Navier-Stokes equations are formulated, are able to smooth out or overcome the ripples, eddies and whorls of turbulence, as mentioned above, the more viscous the fluids the more able they are in doing so. Hence, the more viscous the fluid the more successful the Navier-Stokes equations should be in describing the motion of the fluid.

If turbulence or chaos could be predicted by a mathematical equation or equations then it is not really turbulence or chaos. One should bear in mind that chaos, as the term implies, results in disorder, having no discernable pattern, confusion and puzzlement, which is the contrary of the state of being orderly, having an obvious pattern, being deterministic and being predictable. Nevertheless, whatever method we adopt for describing turbulence or chaos we still need to confirm its validity through experiments, just as the validity of the Navier-Stokes equations for two-dimensional fluid motions has been confirmed by physical experiments. Thus, we should first proceed with the physical experiments to get a better understanding of turbulence or chaos in incompressible fluids.

Fluid Viscosity and Turbulence

The more viscous (sticky or gooey) the incompressible fluid is the less able it is to flow or move freely, the less runny it is, and, as stated above, the more able it is to overcome or smooth out the whorls, eddies and ripples of turbulence. A fluid might be so viscous and its flow so restricted that it appears almost like a solid. Such fluids which come to mind are, e.g., paints and polymer solutions. What does all this imply?

It is evident that there is a threshold or cut-off point in the viscosity of a fluid at and above which turbulence does not affect it much, which means that the effects of turbulence are only marked if a fluid has a viscosity that is less than this threshold or cut-off point. At or beyond this threshold, this cut-off point, one could expect the Navier-Stokes equations not to fare too badly.

How would the introduction of a large initial force affect a viscous fluid whose viscosity is at or above this threshold? If a large initial force is introduced, this fluid could be expected to remain "lumpy" instead of moving more freely or becoming more runny, i.e., turbulence could not be expected to appear.

The opposite is true below this threshold. Below this cut-off point or threshold, i.e., if the fluid is not too viscous, with the initial force large enough, the fluid's Reynolds number stands a chance of exceeding 2000 to 3000, with turbulence setting in, and whorls, eddies and ripples forming in the fluid. The Navier-Stokes equations would encounter difficulties.

Two-Dimensional Flows and their Solutions

Solutions of the Navier-Stokes equations result in velocity vectors, q, and pressures, p, which satisfy both the momentum equations and the continuity equation. If one were given such a combination, [q, p], one could check whether it constitutes a solution by substitution into the equations. How such a solution is found is something else. Any general step leading to this goal is helpful. For two-dimensional flows, it is possible to get rid of the continuity equation from the system of equations by using only functions which satisfy the continuity equation. This elimination is a formal step towards a solution. The functions which affect this elimination are the stream functions.

A flow could be defined as two-dimensional when its description in Cartesian coordinates shows no z-component of the velocity and no dependence on the z-coordinate. A flow like this could be described in the $z = 0$ plane, with the velocity vector and the streamlines lying in this plane. Moreover, the $z = C$ planes, which are parallel to the $z = 0$ plane, display a flow pattern which is identical to that in the $z = 0$ plane. The $z = 0$ plane is called the representative plane.

Three-Dimensional Flows and Practical Solutions

Historically, the dynamics of turbulence or chaos is the corollary of Poincare's "three-body" problem concerning planetary motions which Poincare found to be very complex and unsolvable.

Geometrically, turbulence in fluids cannot be described by the use of the "Poincare section", i.e., there is no periodic solution for it. (In the use of the "Poincare section", for there to be a periodic solution, a circular curve must return to the section at its exact starting point. In the condition of turbulence the curve would not return to its exact starting point and there is therefore no periodic solution.)

Our solutions here are to be implemented as much as possible in the spirit of the Navier-Stokes equations that pertain to the "behaviour" of an incompressible viscous fluid in three dimensions (x, y, z). We aim to overcome the obstacle encountered by the Navier-Stokes equations in the three-dimensional case.

At least two methods of experimentation could be carried out to get a good picture of turbulence in incompressible fluids. This "picture" is however only an "approximation", with a "factor of accuracy" - which comprises an upper limit and a lower limit. The first method involves bouncing laser beams off a reflective strip that is immersed in the turbulent viscous fluid, making use of the Doppler effect, to determine the velocity of the strip, and, thus, the velocity of the fluid. However, this method only provides data obtained from one point, axis or dimension, though we could extrapolate the data for the other two dimensions to get a three-dimensional picture. The second method involves the simulation of fluid motions in conditions of turbulence with powerful computers. The fact that turbulence or chaos is unpredictable implies that the result of fluid motions in turbulent conditions is anybody's guess. The guesses could vary widely in scope but each of them has a "probability" of being correct. A statistical method of analysing fluid motions in conditions of turbulence would be strongly recommended, which is in some way similar in principle to the above-mentioned statistical mechanics. Though statistical mechanics could be used as an example for describing the behaviours of incompressible fluids under conditions of turbulence (e.g., the way it is being used to predict the behaviours and positions of quantum particles - only in a probabilistic manner), applying its methods is bound to face practical difficulties. (How do we assign "probability ratings" to the behaviours of quantum particles? What are the bases for such "ratings"? Are the "ratings" based on experimental proofs or data, which should be the case?) A suggestion is to adopt a method of analysing fluid flows under conditions of turbulence based on statistical data obtained through actual, rigorous experiments, the utilisation of statistical and interpolation/extrapolation methods, and, sound common sense or logic.

The second method, which the author strongly recommends, involves carrying out an experiment through computer simulation. Computer simulation is very powerful and is commonly used today. Instead of carrying out an experiment involving real fluids and real turbulence, which would be cumbersome, we would simulate this experiment with powerful computers, which would be more practicable and would cut down costs and save time considerably. (In computer simulations electrons are actually being utilised to represent the object or objects being simulated and these simulations could be carried out in three dimensions - in this case viscous, incompressible fluid in turbulence with Reynolds numbers above 3000 being simulated in three dimensions as is described below.) As the flow of the fluid is completely turbulent when the Reynolds number exceeds about 2000 to 3000, we could via our powerful computers simulate fluid flows under conditions of turbulence for Reynolds numbers of, e.g., 3050, 3300, 3550, 3800, 4050, 4300, 4550, 4800 and 5050 respectively and thereby obtain the velocity results of fluid motion (by the process of iteration) for each of these Reynolds numbers.

We would use three powerful computers or work stations which work at very high speeds for these computer simulations. For each of the above-mentioned nine Reynolds numbers (3050, 3300, 3550, 3800, 4050, 4300, 4550, 4800 and 5050) the three computers or work stations would carry out the following. The first computer would perform the simulation of fluid movement under conditions of turbulence and the fluid's velocity at a point, P, in the fluid would be measured in the x direction or axis (front view or plane) by a probe at this point, P, and recorded. The second computer would perform the same simulation but at a different plane or view, say, the side view (y), and the fluid's velocity at point P in the fluid would be measured in the y direction or axis (side view or plane) by a probe at point P and recorded. The third computer would perform the same simulation at a yet different plane or view, now, the top view (z), and the fluid's velocity at point P in the fluid would be measured in the z direction or axis (top view or plane) by a probe at point P and recorded. Hence, we now have the fluid's velocities (v = m/s) at each of the three planes or axes, x, y, z, for the same Reynolds number, which are as follows:-

(i)   The fluid's velocity in the x axis - $v(x)$
(ii) The fluid's velocity in the y axis - $v(y)$
(iii) The fluid's velocity in the z axis - $v(z)$

For each of the nine Reynolds numbers, the three computers or work stations would each carry out (iterate) the simulations, say, 1 million times (the more simulations the better) to produce respectively 1 million values for $v(x)$, 1 million values for $v(y)$ and 1 million values for $v(z)$.

(The principles involved in measuring the velocities of the fluid in the x, y, z directions or axes are explained here. The probes mentioned above should be made of thin material and plate-like and should offer minimal resistance against the flow of the fluid, i.e., offer minimal interference with the flow of the fluid. For each axis or plane, the flat surface of the probe should face the direction of the axis. The velocity of the fluid at each axis, plane or dimension is a function of the fluid pressure on the flat surface of the probe, the higher the fluid pressure the higher the velocity of the fluid and vice versa. The probe in each axis or plane should be able to measure both positive (+) velocities and negative (-) velocities, i.e., velocities in the opposite directions.)

This process for all the above-mentioned nine Reynolds numbers would be carried out, giving a total of 27 million simulations, and, 9 million values for $v(x)$, 9 million values for $v(y)$ and 9 million values for $v(z)$.

We next apply the statistical method of time-series analysis (which is a method of statistical analysis designed to eliminate seasonal variations in trends, used for forecasting and prediction, but implemented with some modification here). For each of the above-mentioned nine Reynolds numbers, we compute (with the aid of the computer) the moving averages (m) for the 1 million $v(x)$'s, 1 million $v(y)$'s and 1 million $v(z)$'s, regardless of whether $v(x)$, $v(y)$ and $v(z)$ are positive (+) or negative (-), by only adding (no subtracting) and dividing, as follows:-

(i) $v_1(x) + v_2(x) \div 2 = m_1(x)$
(ii) $v_1(x) + v_2(x) + v_3(x) \div 3 = m_2(x)$
(iii) $v_1(x) + v_2(x) + v_3(x) + v_4(x) \div 4 = m_3(x)$

.
.

(*q) $v_1(x) + v_2(x) + v_3(x) + v_4(x) + \ldots\ldots + v_{q+1}(x) \div q + 1 = m_q(x)$

(*: $q = 999{,}999$)

whereby $m_1(x), m_2(x), m_3(x) \ldots m_q(x)$ are the 999,999 moving averages (m) for v (x).

(The same applies to the 1 million v (y)'s and the 1 million v (z)'s.)

Thus, we have 999,999 (q) moving averages (m) each for each of the 1 million v (x)'s, 1 million v (y)'s and 1 million v (z)'s for each of the above-mentioned nine Reynolds numbers (giving a total of 26,999,973 moving averages (m) for all the above-mentioned nine Reynolds numbers).

For each of the above-mentioned nine Reynolds numbers, substituting 999,999 with q, we get the qth. moving averages (m_q) for v (x), v (y) and v (z), which are actually each respectively the 999,999th. average velocity of the fluid in each of the three dimensions or axes, x, y, z, which are as follows:-

(i) $m_q(x)$
(ii) $m_q(y)$
(iii) $m_q(z)$

Upon the completion of all the computer simulations and computations of the moving averages (m), we would produce a statistical table with the fluid velocities, $m_q(x)$, $m_q(y)$ and $m_q(z)$, in the three dimensions or axes, x (front view), y (side view) and z (top view), for each of the nine Reynolds numbers: 3050, 3300, 3550, 3800, 4050, 4300, 4550, 4800 and 5050. For each of the three velocities, $m_q(x)$, $m_q(y)$ and $m_q(z)$, for each of these nine Reynolds numbers, we would include an upper limit of accuracy, u (x), u (y) and u (z) (each of which is respectively the largest moving average (m) that has been obtained through carrying out the 1 million simulations for each of the three dimensions or axes, x, y, z), and, a lower limit of accuracy, l (x), l (y) and l (z) (each of which is respectively the smallest moving average (m) which has been obtained through carrying out the 1 million simulations for each of the three dimensions or axes, x, y, z). For example, for each of these nine Reynolds numbers we would have the following velocities with their respective upper and lower limits of accuracy:-

(i)   x dimension or axis - $m_q(x)$, u (x), l (x)
(ii)   y dimension or axis - $m_q(y)$, u (y), l (y)
(iii)   z dimension or axis - $m_q(z)$, u (z), l (z)

However, for the Reynolds numbers between these nine Reynolds numbers, i.e., the "intermediate" Reynolds numbers (such as 3200, 4400 and 4950, e.g.), we would obtain the velocity results with their respective upper and lower limits of accuracy, i.e., the m_q's and their respective u's and l's, through interpolation (i.e., estimate the velocity results and their respective upper and lower limits of accuracy - some methods of interpolation are the Lagrange interpolation and the Gregory-Newton interpolation).

With the above-mentioned statistical data for the nine Reynolds numbers we would apply the rules of vector calculus to obtain for each of these nine Reynolds numbers the "resultant" velocity of the fluid in three dimensions or axes (x, y, z) and its upper and lower limits of accuracy. For example, we could obtain the "resultant" velocity and its upper and lower limits of accuracy for unidirectional fluid flow in the three dimensions or axes (x, y, z) as follows:-

$m_q(x, y, z)$, u (x, y, z), l (x, y, z) $= +\, m_q(x) + m_q(y) + m_q(z);\; +\,u(x) + u(y) + u(z);\; +\,l(x) + l(y) + l(z)$

where $m_q(x, y, z)$ is the "resultant" velocity of the three moving averages $m_q(x)$, $m_q(y)$ and $m_q(z)$, whereby $m_q(x)$ would be treated as positive (+) if more than 50% of the 1 million $v(x)$'s are positive (+) and treated as negative (-) if more than 50% of the 1 million $v(x)$'s are negative (-), $m_q(y)$ would be treated as positive (+) if more than 50% of the 1 million $v(y)$'s are positive (+) and treated as negative (-) if more than 50% of the 1 million $v(y)$'s are negative (-), and, $m_q(z)$ would be treated as positive (+) if more than 50% of the 1 million $v(z)$'s are positive (+) and treated as negative (-) if more than 50% of the 1 million $v(z)$'s are negative (-).

Hence, when more than 50% of the 1 million $v(x)$'s, $v(y)$'s or $v(z)$'s are negative (-), i.e., move in the opposite direction, and their moving average (m_q) is hence treated as negative (-), we have to subtract $m_q(x)$, $m_q(y)$ or $m_q(z)$, e.g., if $m_q(z)$ is negative (-), we have to compute the resultant" velocity $m_q(x, y, -z)$ as follows:-

$m_q(x, y, -z) = +m_q(x) + m_q(y) - m_q(z)$, i.e., subtract $m_q(z)$ from $+m_q(x) + m_q(y)$,

with its upper and lower limits of accuracy computed as follows:

(i) $u(x, y, -z) = +u(x) + u(y) - u(z)$, where $u(x)$ is the largest moving average $(m(x))$ among the 999,999 (q) moving averages (m) for the 1 million $v(x)$'s, $u(y)$ is the largest moving average $(m(y))$ among the 999,999 (q) moving averages (m) for the 1 million $v(y)$'s, and, $u(z)$ is the largest moving average $(m(z))$ among the 999,999 (q) moving averages (m) for the 1 million $v(z)$'s.

(ii) $l(x, y, -z) = +l(x) + l(y) - l(z)$, where $l(x)$ is the smallest moving average $(m(x))$ among the 999,999 (q) moving averages (m) for the 1 million $v(x)$'s, $l(y)$ is the smallest moving average $(m(y))$ among the 999,999 (q) moving averages (m) for the 1 million $v(y)$'s, and, $l(z)$ is the smallest moving average $(m(z))$ among the 999,999 (q) moving averages (m) for the 1 million $v(z)$'s.

"Velocity" diagrams (or vector diagrams) could be produced for all the possible "resultant" velocities indicated below, showing their directions of flow, which vary:-

(i) For $+m_q(x) + m_q(y) + m_q(z)$; $+u(x) + u(y) + u(z)$; $+l(x) + l(y) + l(z)$, we get the "resultant" velocity $m_q(x, y, z)$, $u(x, y, z)$, $l(x, y, z)$.

(ii) For $-m_q(x) - m_q(y) - m_q(z)$; $-u(x) - u(y) - u(z)$; $-l(x) - l(y) - l(z)$, we get the "resultant" velocity $m_q(-x, -y, -z)$, $u(-x, -y, -z)$, $l(-x, -y, -z)$.

(iii) For $-m_q(x) - m_q(y) + m_q(z)$; $-u(x) - u(y) + u(z)$; $-l(x) - l(y) + l(z)$, we get the "resultant" velocity $m_q(-x, -y, z)$, $u(-x, -y, z)$, $l(-x, -y, z)$ (where $-m_q(x) - m_q(y) > +m_q(z)$).

(iv) For $+m_q(x) - m_q(y) - m_q(z)$; $+u(x) - u(y) - u(z)$; $+l(x) - l(y) - l(z)$, we get the "resultant" velocity $m_q(x, -y, -z)$, $u(x, -y, -z)$, $l(x, -y, -z)$ (where $-m_q(y) - m_q(z) > +m_q(x)$).

(v) For $-m_q(x) + m_q(y) - m_q(z)$; $-u(x) + u(y) - u(z)$; $-l(x) + l(y) - l(z)$, we get the "resultant" velocity $m_q(-x, y, -z)$, $u(-x, y, -z)$, $l(-x, y, -z)$ (where $-m_q(x) - m_q(z) > +m_q(y)$).

(vi) For $-m_q(x) + m_q(y) + m_q(z)$; $-u(x) + u(y) + u(z)$; $-l(x) + l(y) + l(z)$, we get the "resultant" velocity $m_q(-x, y, z)$, $u(-x, y, z)$, $l(-x, y, z)$ (where $+m_q(y) + m_q(z) > -m_q(x)$).

(vii) For $+m_q(x) - m_q(y) + m_q(z)$; $+u(x) - u(y) + u(z)$; $+l(x) - l(y) + l(z)$, we get the "resultant" velocity $m_q(x, -y, z)$, $u(x, -y, z)$, $l(x, -y, z)$ (where $+m_q(x) + m_q(z) > -m_q(y)$).

(viii) For $+m_q(x) + m_q(y) - m_q(z)$; $+u(x) + u(y) - u(z)$; $+l(x) + l(y) - l(z)$, we get the "resultant" velocity $m_q(x, y, -z)$, $u(x, y, -z)$, $l(x, y, -z)$ (where $+m_q(x) + m_q(y) > -m_q(z)$).

So far, in the above "velocity" scenarios, for those with either, one negative velocity ($-m_q$) and two positive velocities ($+m_q$), or, one positive velocity ($+m_q$) and two negative velocities ($-m_q$), it is assumed that for the former the two positive velocities ($+m_q$) combined together are larger than the one negative velocity ($-m_q$), and, for the latter it is assumed that the two negative velocities ($-m_q$) combined together are larger than the one positive velocity ($+m_q$). Now, what happens when for the former, the two positive velocities ($+m_q$) combined together are equal to the one negative velocity ($-m_q$), and, for the latter, the two negative velocities ($-m_q$) combined together are equal to the one positive velocity ($+m_q$)? In each of these cases, the "resultant" velocity ($m_q(x, y, z)$, $u(x, y, z)$, $l(x, y, z)$) would be equal to zero, i.e., there would be no "resultant" velocity.

What happens when the negative velocity, $-m_q(x)$, $-m_q(y)$, or, $-m_q(z)$, is larger than the other two positive velocities combined together, e.g., $-m_q(x) > +m_q(y) + m_q(z)$, $-m_q(y) > +m_q(x) + m_q(z)$, or, $-m_q(z) > +m_q(x) + m_q(y)$? For such cases we have the following "resultant" velocities, whose directions of flow differ from each other:-

(i) For $-m_q(x) + m_q(y) + m_q(z)$; $-u(x) + u(y) + u(z)$; $-l(x) + l(y) + l(z)$, where $-m_q(x) > +m_q(y) + m_q(z)$, we get the "resultant" velocity $m_q(-x, y, z)$, $u(-x, y, z)$, $l(-x, y, z)$.

(ii) For $+m_q(x) - m_q(y) + m_q(z)$; $+u(x) - u(y) + u(z)$; $+l(x) - l(y) + l(z)$, where $-m_q(y) > +m_q(x) + m_q(z)$, we get the "resultant" velocity $m_q(x, -y, z)$, $u(x, -y, z)$, $l(x, -y, z)$.

(iii) For $+m_q(x) + m_q(y) - m_q(z)$; $+u(x) + u(y) - u(z)$; $+l(x) + l(y) - l(z)$, where $-m_q(z) > +m_q(x) + m_q(y)$, we get the "resultant" velocity $m_q(x, y, -z)$, $u(x, y, -z)$, $l(x, y, -z)$.

What happens then when the positive velocity, $+m_q(x)$, $+m_q(y)$, or $+m_q(z)$ is larger than the other two negative velocities combined together, e.g., $+m_q(x) > -m_q(y) - m_q(z)$, $+m_q(y) > -m_q(x) - m_q(z)$, or, $+m_q(z) > -m_q(x) - m_q(y)$? For such cases we have the following "resultant" velocities, whose directions of flow differ from each other:-

(i) For $+m_q(x) - m_q(y) - m_q(z)$; $+u(x) - u(y) - u(z)$; $+l(x) - l(y) - l(z)$, where $+m_q(x) > -m_q(y) - m_q(z)$, we get the "resultant" velocity $m_q(x, -y, -z)$, $u(x, -y, -z)$, $l(x, -y, -z)$.

(ii) For $-m_q(x) + m_q(y) - m_q(z)$; $-u(x) + u(y) - u(z)$; $-l(x) + l(y) - l(z)$, where $+m_q(y) > -m_q(x) - m_q(z)$, we get the "resultant" velocity $m_q(-x, y, -z)$, $u(-x, y, -z)$, $l(-x, y, -z)$.

(iii) For $-m_q(x) - m_q(y) + m_q(z)$; $-u(x) - u(y) + u(z)$; $-l(x) - l(y) + l(z)$, where $+m_q(z) > -m_q(x) - m_q(y)$, we get the "resultant" velocity $m_q(-x, -y, z)$, $u(-x, -y, z)$, $l(-x, -y, z)$.

For the three velocities in the x, y, z dimensions or axes, i.e., $m_q(x)$, $m_q(y)$ and $m_q(z)$, there are 14 possible "resultant" velocities, excluding the null "resultant" velocities, of which there are six possible cases (which are not expected to be likely to occur in turbulence), as shown above. Each of these "resultant" velocities would have an upper limit of accuracy ($u(x, y, z)$) and a lower limit of accuracy ($l(x, y, z)$). For fluid velocities ($m_q(x)$, $m_q(y)$, or, $m_q(z)$) in the opposite direction their values are negative. We could expect turbulence to be characterized by any of these 14 "resultant" velocities.

Nevertheless, there is a possibility that one or more of the velocities, $m_q(x)$, $m_q(y)$ and $m_q(z)$, might be equal to zero, though the chances of this occurring in turbulence might be remote:-

(i) If, e.g., $m_q(z) = 0$, then $m_q(x, y, z = 0)$, $u(x, y, z = 0)$, $l(x, y, z = 0)$ would be a "resultant" velocity in two dimensions or axes only, with the fluid moving in the two dimensions or axes, x and y, only.

(ii) For $m_q(x) = 0$, we get the "resultant" velocity $m_q(x = 0, y, z)$, $u(x = 0, y, z)$, $l(x = 0, y, z)$.

(iii) For $m_q(y) = 0$, we get the "resultant" velocity $m_q(x, y = 0, z)$, $u(x, y = 0, z)$, $l(x, y = 0, z)$.

As in the three-dimensional case above, there is the possibility in the two-dimensional case that one or both of the two velocities are negative or positive velocities. If, in the case whereby one of the two velocities is negative while the other is positive, the negative velocity is equal to the positive velocity (of which there are six possible cases), then the "resultant" velocity is null (zero). The following are the possible "resultant" velocities:-

(i) For $m_q(z) = 0$, we get the "resultant" velocity $m_q(x, -y, z = 0)$, $u(x, -y, z = 0)$, $l(x, -y, z = 0)$, when $m_q(x)$ is positive while $m_q(y)$ is negative, whereby $-m_q(y) > +m_q(x)$.

(ii) For $m_q(z) = 0$, we get the "resultant" velocity $m_q(x, -y, z = 0)$, $u(x, -y, z = 0)$, $l(x, -y, z = 0)$, when $m_q(x)$ is positive while $m_q(y)$ is negative, whereby $+m_q(x) > -m_q(y)$.

(iii) For $m_q(z) = 0$, we get the "resultant" velocity $m_q(-x, y, z = 0)$, $u(-x, y, z = 0)$, $l(-x, y, z = 0)$, when $m_q(y)$ is positive while $m_q(x)$ is negative, whereby $-m_q(x) > +m_q(y)$.

(iv) For $m_q(z) = 0$, we get the "resultant" velocity $m_q(-x, y, z = 0)$, $u(-x, y, z = 0)$, $l(-x, y, z = 0)$, when $m_q(y)$ is positive while $m_q(x)$ is negative, whereby $+m_q(y) > -m_q(x)$.

(v) For $m_q(z) = 0$, we get the "resultant" velocity $m_q(-x, -y, z = 0)$, $u(-x, -y, z = 0)$, $l(-x, -y, z = 0)$, when $m_q(x)$ and $m_q(y)$ are both negative.

(vi) For $m_q(z) = 0$, we get the "resultant" velocity $m_q(x, y, z = 0)$, $u(x, y, z = 0)$, $l(x, y, z = 0)$, when $m_q(x)$ and $m_q(y)$ are both positive.

(vii) For $m_q(x) = 0$, we get the "resultant" velocity $m_q(x = 0, -y, z)$, $u(x = 0, -y, z)$, $l(x = 0, -y, z)$, when $m_q(z)$ is positive while $m_q(y)$ is negative, whereby $-m_q(y) > +m_q(z)$.

(viii) For $m_q(x) = 0$, we get the "resultant" velocity $m_q(x = 0, -y, z)$, $u(x = 0, -y, z)$, $l(x = 0, -y, z)$, when $m_q(z)$ is positive while $m_q(y)$ is negative, whereby $+m_q(z) > -m_q(y)$.

(ix) For $m_q(x) = 0$, we get the "resultant" velocity $m_q(x = 0, y, -z)$, $u(x = 0, y, -z)$, $l(x = 0, y, -z)$, when $m_q(y)$ is positive while $m_q(z)$ is negative, whereby $-m_q(z) > +m_q(y)$.

(x) For $m_q(x) = 0$, we get the "resultant" velocity $m_q(x=0, y, -z)$, $u(x=0, y, -z)$, $l(x=0, y, -z)$, when $m_q(y)$ is positive while $m_q(z)$ is negative, whereby $+ m_q(y) > - m_q(z)$.

(xi) For $m_q(x) = 0$, we get the "resultant" velocity $m_q(x=0, -y, -z)$, $u(x=0, -y, -z)$, $l(x=0, -y, -z)$, when $m_q(y)$ and $m_q(z)$ are both negative.

(xii) For $m_q(x) = 0$, we get the "resultant" velocity $m_q(x=0, y, z)$, $u(x=0, y, z)$, $l(x=0, y, z)$, when $m_q(y)$ and $m_q(z)$ are both positive.

(xiii) For $m_q(y) = 0$, we get the "resultant" velocity $m_q(-x, y=0, z)$, $u(-x, y=0, z)$, $l(-x, y=0, z)$, when $m_q(z)$ is positive while $m_q(x)$ is negative, whereby $- m_q(x) > + m_q(z)$.

(xiv) For $m_q(y) = 0$, we get the "resultant" velocity $m_q(-x, y=0, z)$, $u(-x, y=0, z)$, $l(-x, y=0, z)$, when $m_q(z)$ is positive while $m_q(x)$ is negative, whereby $+ m_q(z) > - m_q(x)$.

(xv) For $m_q(y) = 0$, we get the "resultant" velocity $m_q(x, y=0, -z)$, $u(x, y=0, -z)$, $l(x, y=0, -z)$, when $m_q(x)$ is positive while $m_q(z)$ is negative, whereby $- m_q(z) > + m_q(x)$.

(xvi) For $m_q(y) = 0$, we get the "resultant" velocity $m_q(x, y=0, -z)$, $u(x, y=0, -z)$, $l(x, y=0, -z)$, when $m_q(x)$ is positive while $m_q(z)$ is negative, whereby $+ m_q(x) > - m_q(z)$.

(xvii) For $m_q(y) = 0$, we get the "resultant" velocity $m_q(-x, y=0, -z)$, $u(-x, y=0, -z)$, $l(-x, y=0, -z)$, when $m_q(x)$ and $m_q(z)$ are both negative.

(xviii) For $m_q(y) = 0$, we get the "resultant" velocity $m_q(x, y=0, z)$, $u(x, y=0, z)$, $l(x, y=0, z)$, when $m_q(x)$ and $m_q(z)$ are both positive.

(xix) For $m_q(x) = 0$ and $m_q(y) = 0$, we get the "resultant" velocity $m_q(x=0, y=0, z)$, $u(x=0, y=0, z)$, $l(x=0, y=0, z)$ in one dimension or axis, which is positive.

(xx) For $m_q(x) = 0$ and $m_q(y) = 0$, we get the "resultant" velocity $m_q(x=0, y=0, -z)$, $u(x=0, y=0, -z)$, $l(x=0, y=0, -z)$ in one dimension or axis, which is negative.

(xxi) For $m_q(x) = 0$ and $m_q(z) = 0$, we get the "resultant" velocity $m_q(x=0, y, z=0)$, $u(x=0, y, z=0)$, $l(x=0, y, z=0)$ in one dimension or axis, which is positive.

(xxii) For $m_q(x) = 0$ and $m_q(z) = 0$, we get the "resultant" velocity $m_q(x=0, -y, z=0)$, $u(x=0, -y, z=0)$, $l(x=0, -y, z=0)$ in one dimension or axis, which is negative.

(xxiii) For $m_q(y) = 0$ and $m_q(z) = 0$, we get the "resultant" velocity $m_q(x, y=0, z=0)$, $u(x, y=0, z=0)$, $l(x, y=0, z=0)$ in one dimension or axis, which is positive.

(xxiv) For $m_q(y) = 0$ and $m_q(z) = 0$, we get the "resultant" velocity $m_q(-x, y=0, z=0)$, $u(-x, y=0, z=0)$, $l(-x, y=0, z=0)$ in one dimension or axis, which is negative.

After all the simulations and computations of moving averages (m) for the 27 million fluid velocities (v) in the x, y, z dimensions or axes, we firstly obtain a table, which would act as a rough statistical guide for fluid flow under conditions of turbulence, with the following statistical data, for each of the nine Reynolds numbers (3050, 3300, 3550, 3800, 4050, 4300, 4550, 4800 and 5050):-

(1) (i) The fluid's average velocity in the x dimension or axis - $m_q(x)$

       (ii) This fluid average velocity's upper limit of accuracy - u (x)   (i.e., largest moving average (m) for v (x))
       (iii) This fluid average velocity's lower limit of accuracy - l (x)   (i.e., smallest moving average (m) for v (x))

(2) (i) The fluid's average velocity in the y dimension or axis - m_q (y)
     (ii) This fluid average velocity's upper limit of accuracy - u (y)   (i.e., largest moving average (m) for v (y))
     (iii) This fluid average velocity's lower limit of accuracy - l (y)   (i.e., smallest moving average (m) for v (y))

(3) (i) The fluid's average velocity in the z dimension or axis - m_q (z)
     (ii) This fluid average velocity's upper limit of accuracy - u (z)   (i.e., largest moving average (m) for v (z))
     (iii) This fluid average velocity's lower limit of accuracy - l (z)   (i.e., smallest moving average (m) for v (z))

(4) (i) The fluid's "resultant" velocity in the x, y, z dimensions or axes - m_q (x, y, z) (i.e., +/- m_q (x) +/- m_q (y) +/- m_q (z))
     (ii) This fluid "resultant" velocity's upper limit of accuracy - u (x, y, z) (i.e., +/- m (x) +/- m (y) +/- m (z))
     (iii) This fluid "resultant" velocity's lower limit of accuracy - l (x, y, z) (i.e., +/- m (x) +/- m (y) +/- m (z))

The "velocity" diagrams depicting the various "resultant" velocities and the directions of flow should also be included, a total of 27 diagrams, three "velocity" diagrams for each of the nine Reynolds numbers - one for m_q (x, y, z), one for u (x, y, z) and one for l (x, y, z). This table may be titled "Statistical Data Of An Incompressible Fluid's Velocities For The Nine Reynolds Numbers: 3050, 3300, 3550, 3800, 4050, 4300, 4550, 4800 And 5050".

For the "intermediate" Reynolds numbers, e.g., 3200, 4400 and 4950, we would interpolate the fluid's respective velocities and their upper and lower limits of accuracy with the above-mentioned statistical table (Statistical Data Of An Incompressible Fluid's Velocities For The Nine Reynolds Numbers: 3050, 3300, 3550, 3800, 4050, 4300, 4550, 4800 And 5050) and the complementary table (Velocity Results Of The 27 Million Simulations In The X, Y, Z Dimensions) described below as guides, i.e., estimate them - some methods of interpolation that could be adopted are the Lagrange interpolation and the Gregory-Newton interpolation. For velocities for Reynolds numbers outside the range of the above nine Reynolds numbers (3050 to 5050), e.g., for Reynolds numbers 2500 and 5200, we would extrapolate them, including their upper and lower limits of accuracy, using the same tables as guides.

With the above techniques (and with proper simulations) we thus have a statistical table of fluid flow velocity results (Statistical Data Of An Incompressible Fluid's Velocities For The Nine Reynolds Numbers: 3050, 3300, 3550, 3800, 4050, 4300, 4550, 4800 And 5050) under conditions of turbulence (with their respective ranges of possibilities or probabilities, i.e., their respective upper limits and lower limits of accuracy) for the various Reynolds numbers which are more or less rigorously based on actual experiments (simulations in this case), a table which is somewhat like, e.g., the statistical tables for t-distribution and chi-squared distribution. In the above-mentioned experiments (or simulations), we could expect more accurate results with more simulations being carried out (more data being collected), the more the simulations carried out the more accurate the results are likely to be. This represents a practical, logical way of roughly approximating the velocities of fluid motions under conditions of turbulence (in three dimensions) at the various Reynolds numbers. Given a fluid velocity (m_q (x, y, z), u (x, y, z), l (x, y, z)) from the above-mentioned statistical table, we could in principle tell (or predict) for a particular Reynolds number the position (position after distance travelled, d_2, relative to the position at d_1 = 0) of an object, e.g., an aluminium strip, which could be taken to represent a section of the fluid, carried along by the fluid in motion, at a point of time, $t_2 > 0$, with the initial time being $t_1 = 0$, or, the distance ($\{d_2 > 0\}$ - $\{d_1 = 0\}$) this object travelled at a point of time, $t_2 > 0$, with the initial time being $t_1 = 0$, by computation with the equation: $d_2 - d_1 = (m_q(x, y, z), u(x, y, z), l(x, y, z)) \times (t_2 - t_1)$. However, in this instance, wherein turbulence rules, this prediction of position or distance is not expected to be accurate (but should be regarded only as a rough approximation) and should be subject to the statistical rule of "probability" - in this case the upper limit ($\{u(x, y, z)\} \times \{t_2 - t_1\}$) and the lower limit ($\{l(x, y, z)\} \times \{t_2 - t_1\}$) of the accuracy of the result obtained, as indicated by the above-mentioned statistical table. The "velocity" diagrams in the statistical table would act as a guide with regards to the positioning of the object and the direction it would travel. Here the movement of the object, which is the aluminium strip, in effect represents the movement of the fluid which carries it along. However, the data reflected in this first statistical table (Statistical Data Of An Incompressible Fluid's Velocities For The Nine Reynolds Numbers: 3050, 3300, 3550, 3800, 4050, 4300, 4550, 4800 And 5050) should be regarded only as a rough approximation. With the further assistance of the complementary table (Velocity Results Of The 27 Million Simulations In The X, Y, Z Dimensions) described below this approximation could be refined.

With the data from the above-mentioned statistical table (Statistical Data Of An Incompressible Fluid's Velocities For The Nine Reynolds Numbers: 3050, 3300, 3550, 3800, 4050, 4300, 4550, 4800 And 5050) it is possible to plot a curve for the fluid flow velocity results ($m_q(x, y, z)$) for the above-mentioned nine Reynolds numbers. If this curve is smooth and linear, which is unlikely, from the gradient of the slope of this curve the differential equations for this curve could be derived, whereby forecasts or predictions would be possible. However, if this curve is rough and nonlinear, which is likely to be the case, then these differential equations would not be obtainable, and, we would have to rely on the above-mentioned table of fluid flow velocity results (Statistical Data Of An Incompressible Fluid's Velocities For The Nine Reynolds Numbers: 3050, 3300, 3550, 3800, 4050, 4300, 4550, 4800 And 5050), and, the complementary table (Velocity Results Of The 27 Million Simulations In The X, Y, Z Dimensions) described below, as statistical guides for our approximation (including interpolation and extrapolation) of velocity results for fluid flows at the various Reynolds numbers. Both these statistical tables with their "velocity" diagrams should complement one another and should be a great help for this approximation process, the more copious the data available there the more effective the approximation should be. It is thus possible to approximate not only fluid velocities but directions of fluid motions as well (with the aid of the "velocity" diagrams). We could continue to plot graphs or curves with these statistical data and attempt to interpolate or extrapolate with them, looking out for trends or patterns, and arrive at some forecasts or predictions, which is expected to be a difficult task. By doing this we could at least get a "feel" or intuitive understanding of the whole situation, which should make the job of forecasting or predicting the outcome easier, an evidently challenging undertaking.

The above-mentioned statistical table, Statistical Data Of An Incompressible Fluid's Velocities For The Nine Reynolds Numbers: 3050, 3300, 3550, 3800, 4050, 4300, 4550, 4800 And 5050, presents the average velocities for the nine Reynolds numbers. These average velocities each represents the velocity "trend" for each of the nine Reynolds numbers. However, carrying out the approximation for the respective velocities for the respective Reynolds numbers, wherein complete turbulence is involved, might not be that easy. It should be remembered that the Navier-Stokes equations fare very badly when turbulence sets in, when a viscous, incompressible fluid behaves in an unpredictable, irregular, chaotic, and, "nonlinear" manner. To complement the first statistical table (Statistical Data Of An Incompressible Fluid's Velocities For The Nine Reynolds Numbers: 3050, 3300, 3550, 3800, 4050, 4300, 4550, 4800 And 5050), it would be appropriate to have a proper tabulation of all the 27 million fluid velocities (v) in the x, y, z dimensions or axes obtained through the simulations in another table (or booklet), which may be titled "Velocity Results Of The 27 Million Simulations In The X, Y, Z Dimensions," which shows whole ranges of velocities (minimum velocity, maximum velocity, and, all the velocities between the minimum and the maximum, a total of 27 million velocities (v's), three million velocities (v's) for each of the nine Reynolds numbers), with their order of listing in this table (or booklet) exactly the same as the order they appeared in during the simulations, with indications whether they are positive (+) or negative (-) velocities, and, the 999,999

(q) moving averages (m) each for each of the 1 million v (x)'s, 1 million v (y)'s and 1 million v (z)'s for each of the nine Reynolds numbers (giving a total of 26,999,973 moving averages (m) for all the nine Reynolds numbers) included. The "velocity" diagrams described above, which would also be helpful as guides, should be incorporated, e.g., as an appendix, in this table (or booklet). With this additional table and its "velocity" diagrams acting as a further guide, the task of approximating (including interpolating and extrapolating) the velocities for the various Reynolds numbers (for turbulence) would be made easier. With these two tables there are now more or less solid and realistic data to carry out the approximation (including interpolation and extrapolation) - for both fluid velocities, and, directions of fluid motions. The approximations could also include probabilities of occurrence (an example of which is presented below). However, sound common sense, good intuition and a sharp eye for patterns and details are important for the approximation (including interpolation and extrapolation). Since turbulence is a much complex phenomenon, the approximation (including interpolation and extrapolation) should be carried out with great care and patience. There are two choices here now. We could supplant the approximation method of the Navier-Stokes equations with the above-mentioned statistical method. Alternatively, we could use this statistical method as a complement to the Navier-Stokes equations, whereby we might have the "best of both worlds".

Conclusion

There is a possibility, however remote, that turbulence or chaos when viewed en masse, on a very large scale, would appear to be smooth, present some sort of pattern or appear to have some order. According to the precepts of fractal geometry, a relatively new branch of mathematics pioneered by Benoit Mandelbrot, phenomena which appear random, when viewed en masse, display some orderliness and pattern, which could be termed "fractal". Thus, a new mathematical technique for describing the flow of an incompressible viscous fluid in three dimensions, i.e., in turbulence, a probably statistical one involving large samples of data, like the method described just above, is a logical step.

As for the case of computer simulation, it is becoming more and more popular. Simulation has been used in the physical sciences, as well as the engineering sciences, e.g., in aeronautical design, electronic circuit design and mechanical design. The author himself has considerable experience with computer simulation. Simulation has generally proven to be cost-saving, time-saving and effective. In engineering applications, e.g., simulation has made it unnecessary to produce or manufacture the prototypes for any new product designs for the purpose of feasibility studies, which could be costly affairs. The feasibility studies are now carried out directly through the simulation exercises, whereby it is possible to quickly find out whether the designs would work or not. The only serious obstacle to simulation appears to be the cost of the simulation software itself, which could come up to many thousands of dollars, so that a cost/benefit analysis for using the software is necessary. The benefits and cost-savings for using the software should outweigh the cost of the software itself in order for its use to be justifiable. The other problem might be a technical one; some simulation software are complex and not user-friendly, requiring a long learning-curve, and these are usually the more powerful software with more built-in features. Nevertheless, once these software have been mastered their use would bring about relatively greater and more effective results. Some might feel that such powerful, and hence more expensive, software are an overkill and prefer to go for something at the lower end which is also cheaper. Such software usually have powerful features such as allowing one to have "walk-through" views and "inside-out" views of an object which would be physically impossible otherwise. Such are the great powers of simulation software now. The simulation of turbulence with powerful software and computers would certainly prove to be very useful.

As nonlinear equations such as the Navier-Stokes equations have to rely on approximation and exact solutions are highly unlikely, especially for turbulent fluid movement, a statistical guide based on actual data collected such as the guides or tables described just above is definitely a boon. A statistical guide that is based on reliable data collected and which has been put to the test and fine-tuned would be in a better position to lead us to more accurate approximations than the Navier-Stokes equations.

From the data in the above-mentioned statistical tables it is now possible, e.g., to make approximations or estimates of various fluid velocities (as well as the directions of various fluid motions) with various probabilities of occurrence for the various Reynolds numbers, such as the following:-

1) x metres/second (accompanied by the appropriate "velocity" diagram): a percent probability of occurrence
2) y metres/second (accompanied by the appropriate "velocity" diagram): b percent probability of occurrence
3) z metres/second (accompanied by the appropriate "velocity" diagram): c percent probability of occurrence

Etc.

The Navier-Stokes equations do not have any allowance for such probabilistic approximations. These partial differential equations have been found to be solvable for the two-dimensional case, i.e., for each of the equations a function (or, solution) could be found that, when substituted for the dependent variable in the equation, leads to an identity. But for three dimensions finding such functions (or, solutions) has been a problem. In other words, though the formulas could be found to describe a two-dimensional fluid motion, such formulas are not available or obtainable for the three-dimensional case, e.g., the case for turbulent fluid motion. Evidently, the geometry of a three-dimensional fluid motion is rather complex, while finding the formulas to describe this complex three-dimensional movement of fluid is a difficult task. To understand this difficulty, consider the motion of a speck of particle carried along in a flowing fluid. The speck could be thrust first in one direction, then another, and another, and so on, sometimes moving in a fairly straight line, other times spiralling around as the current takes it along. This movement of the speck is also the movement of the fluid which carries it along. It is a three-dimensional movement that appears chaotic or turbulent - it is rather complicated and, therefore, not amenable to a description by some formulas. The Navier-Stokes equation makes use of differentiation to obtain the rate of change of some changing quantity, and, in order to do this the value or position or path of that quantity has to be given by an appropriate formula. Differentiation then acts upon this formula to produce another formula which gives the rate of change. Since the formulas for the three-dimensional case, e.g., turbulent fluid motion, are unobtainable the only recourse is approximation. Is there any hope of discovering these formulas in the future? The author is much pessimistic. Let us look again at the case of turbulent fluid motion, an essentially three-dimensional phenomenon. As stated earlier, a curve describing a fluid's movement under conditions of turbulence could be expected to be rough and nonlinear thus making it not possible to derive the differential equations (or formulas) for describing this curve, making predictions relating to the fluid's movement very difficult if not impossible. Moreover, turbulence or chaos implies disorder, irregularity, lack of discernable pattern, confusion and puzzlement. This implies that turbulence could never be described by formulas or differential equations, and, to be able to obtain the formulas or differential equations for turbulence and thus be able to make predictions relating to turbulence would mean that the so-called turbulence is not really turbulence at all, a contradiction. (Do the editors of our dictionaries then have to revise the meanings for turbulence and chaos? According to the Encyclopedic World Dictionary published by Paul Hamlyn, "turbulence" could be defined as "the haphazard secondary motion due to eddies within a moving fluid", or, "irregular motion of the atmosphere, as that indicated by gusts and lulls in the wind", "turbulent flow" is "fluid flow in which the motion at any point varies rapidly in magnitude and direction", and, "chaos" could be defined as "utter confusion or disorder, wholly without organisation or order", or, "the infinity of space or formless matter supposed to have preceded the existence of the ordered universe".) It is thus absurd for us to expect the Navier-Stokes equations to have the solution for a three-dimensional phenomenon such as turbulent or chaotic fluid motion. It is hence appropriate to have another mathematical technique to deal with this difficult situation instead, viz., the statistical method, taking into consideration the example of statistical mechanics in quantum theory. As the outcomes of turbulence or chaos could never be predicted with certainty and the predictions might not even be highly accurate, it is logical and practical to consider the possible outcomes of this phenomenon in a statistical or probabilistic way. Much attempts have already been carried out to understand turbulence or chaos, which is now a hot subject. The author thinks that even if we understand the causes and mechanics or physics of turbulence or chaos it would be naïve to believe it is possible to obtain the formulas for describing such a phenomenon, for, to say that we know how an object which causes confusion and puzzlement (a chaotic object) would behave is to say that we are not confused and puzzled by this object, a self-contradiction. As described above, the logistic equation has been a popular model for chaos but, here again, it would be naïve, and, self-contradicting, to believe that the logistic equation is a sufficient formula for making predictions relating to chaos. Some others might state that only when we could not predict the outcomes of a phenomenon could we regard that phenomenon as chaotic but once we could get those predictions the phenomenon is no more chaotic, which would imply that chaos is transitional and subjective. These are possibly the ones who believe that there is a solution for the Navier-Stokes equations in the three-dimensional case. Simply put, if chaos is predictable it is not chaos. Only when it is really unpredictable could it be chaos. Mathematically, and, objectively speaking too, it has never been found possible to derive the differential equations to describe a nonlinear phenomenon such as turbulence, as there is no regular pattern found in turbulence (which is fluid flow in which the motion at any point varies rapidly in magnitude and direction), and, mathematics, which, in a sense, is the study and analysis of patterns, simply found nothing possible to study or analyse in turbulence since it does not display any discernable, set pattern or regularity, except for the presence of eddies, ripples and whorls, which in the terms of fractal geometry could be described as a fractal characteristic. The only plausible solution appears to be a statistical one, wherein there is some hope of discovering some meaningful patterns or orderly features when large samples of data are analysed. (According to the precepts of fractal geometry, phenomena which appear random when viewed en masse display some orderliness and pattern which could be regarded as a fractal characteristic.) For example, the prime numbers are very random and haphazard entities, yet, when viewed en masse they display a regularity in the way they thin out, whereby it is affirmed that the number of primes not exceeding a given natural number n is approximately $n/\ln n$, in the sense that the ratio of the number of such primes to $n/\ln n$ eventually approaches 1 as n becomes larger and larger, $\ln n$ being the natural logarithm (to the base e) of n (vide the prime number theorem proved in 1896 by Hadamard and Vallee-Poussin).

With statistical methods, such as that described above, it is now possible to evaluate turbulence on a probabilistic basis, which is a more practical and realistic way of looking at turbulence, whose outcomes are uncertain, irregular, haphazard and very difficult if not impossible to predict. As the Navier-Stokes equations fare very poorly in the three-dimensional case, i.e., in the case of turbulence, a new mathematical technique for making approximations for the three-dimensional case, such as the statistical one described above, should be given a chance to take over, to supplant the Navier-Stokes equations, or, at least, to complement them.

After all, the results of the differential equations, such as the Navier-Stokes equations, would still have to be confirmed by actual physical data, or, physical experiment. The more direct, faster or more efficient method of understanding and making forecasts pertaining to the various levels of turbulence or chaos is evidently to execute a well-planned computer simulation exercise (or, at least, a well-planned physical experiment) and apply the proper statistical technique in the interpretation of the data which are obtained through the computer simulation exercise. Such a procedure has been described in detail above.

We should be realistic and refrain from expecting certainly true or almost certainly true forecasts from our "mathematisation" of turbulence or chaos. It is more reasonable or realistic to expect forecasts with only some degree of probability of being true, and, where the forecasts do indeed turn out to be totally true on occasions they should be regarded as rare phenomena not unlike the cases of some fortunate persons hitting the lotteries. In fact, if the day ever arrives when an equation is available for accurately forecasting the outcome of turbulence or chaos, turbulence or chaos would be obsolete. With the lack of a discernable pattern or patterns in turbulence or chaos, such an equation evidently would never be found. In other words, turbulence or chaos, due to its excessively nonlinear, excessively random or patternless, nature, would be practically impossible to describe geometrically through any mathematical equation, e.g., the Navier-Stokes differential equation. A sort of fixed, geometrical shape or pattern has to be discernable to be describable by a mathematical equation. However, in the case of turbulence or chaos, the object in turbulence evidently takes on "different geometrical shapes all the time". Differential equations such as the Navier-Stokes equation are thus bound to fail in helping us to understand turbulence. On the other hand, statistical methods evidently have more chance of success; no matter how slim the chance of success with statistical methods might be, it is evidently still better than that of other mathematical methods. It should be remembered that in the case of statistical methods, the greater the sample size utilised is the greater the chance of success would be. They are therefore the best hope for success for understanding turbulence.

Appendix

The Self-Similarity Concept and Fractal Geometry

The formulation of the self-similarity concept has brought fame to Mitchell Feigenbaum, who had worked in the Los Alamos Laboratory in the early 1970s. This concept, upon which the method of renormalisation in perturbation theory is based, postulates that there is a tendency of identical mathematical structures to recur on many levels. Within a given structure, there would be smaller copies of the same structure, their sizes being determined by the scaling factor. Feigenbaum found that at the utmost tips of the fig-tree, there is some mathematical structure which remains the same when its size is changed (enlarged) by a scaling factor of 4.669, which is found to be a constant like pi (3.142). This structure is the shape of the fig-tree itself. In other words, little whorls could be found within big whorls. Renormalisation has been a well-established technique in chaos theory/fractal geometry and is a mathematical trick which functions rather like a microscope, zooming in on the self-similar structure, removing any approximations, and filtering out everything else. All this shows the universality of some features of chaos. That is, some kind of order or pattern could be found in or is inherent in disorder, turbulence or chaos.

In Feigenbaum's famous fig-tree example, for instance, there is a self-similar mathematical pattern or structure (which is the shape of the fig-tree itself) in the various parts of the fig-tree, i.e., its trunk to bough section, bough to branch section, branch to twig section and twig to twiglet section. Such self-similar mathematical pattern or structure, or, fractal characteristic, could also be found in other aspects of nature, e.g., waves, turbulence or chaos, the structures of viruses and bacteria, polymers and ceramic materials, the universe and many others, even the movements of prices in financial markets, the growths of populations, the sound of music, the flow of blood through our circulatory system, the behaviour of people en masse, etc., which have all spawned a relatively new and important branch of mathematics with wide practical applications known as fractal geometry, which has been pioneered by Benoit Mandelbrot. As a matter of fact, self-similarity or fractal characteristic could be regarded as the fundamental mathematical aspect found in practically everything in nature, and, this new branch of mathematics, fractal geometry, besides having a great practical impact on us also gives us a more profound vision of the universe in which we live and our place in it.

5

CALCULUS AND DIFFERENTIAL EQUATIONS

The motion of fluids which are incompressible could be described by the Navier-Stokes differential equations. Although they are relatively simple-looking, the three-dimensional Navier-Stokes equations misbehave very badly. Even with nice, smooth, reasonably harmless initial conditions, the solutions could wind up being extremely unstable. The field of fluid mechanics would be dramatically altered through a mathematical understanding of the outrageous behaviour of these equations. An explanation why the three-dimensional Navier-Stokes equations are not solvable, i.e., the equations cannot be used to model turbulence or chaos (which is a three-dimensional phenomenon), would be provided.

Sir George Stokes obtained the general equations of motion for a viscous fluid in 1845. The fundamental equation (in vectorial form) governing the flow of a viscous fluid is as follows:-

$$\frac{\partial v}{\partial t} + (v \cdot \nabla)v = -\frac{1}{\rho}\nabla Pe - \nabla \varphi + \frac{\eta}{\rho}\nabla^2 v,$$

where v represents the velocity of the fluid as a function of position, Pe the pressure, φ the gravitational potential, p the density and η the viscosity.

The scientist normally utilises the Navier-Stokes equations as a model to make a forecast of the outcome of a flow. However, for the case of turbulence or chaos making this forecast would be very difficult, if it could be done at all. If turbulence or chaos could be predicted, forecasted or modeled by the Navier-Stokes equations it is by definition not turbulence or chaos, as turbulence or chaos implies lack of predictability, lack of pattern or order and puzzlement. As turbulent flows are three-dimensional, nonlinear and highly unsteady, it would be practically impossible for the Navier-Stokes equations to model them.

Calculus, of which the Navier-Stokes equations are an example, relies on continuity, which is also a common human assumption. If one sees a person running at one moment here and a half-hour later there, one normally assumes the person has run a continuous line covering all the ground in between. It does not occur to him that the person might have stopped to rest or had even hitched a ride, i.e., there had been discontinuities in the path covered by the runner. Calculus, which was co-invented by Isaac Newton and Gottfried von Leibniz and was the greatest innovation of seventeenth-century mathematics, was designed to study continuous change; Leibniz believed deeply in what he described a "principle of continuity". Economists also normally assume continuity when using calculus to model the economy. Continuity is also a fundamental assumption of conventional finance, e.g., the financial mathematics of Batchelier, Black-Scholes, Sharpe and Markowitz all assume continuous change from one price to the next, without which their formulae would not work. However, this assumption is wrong and thus is their mathematics.

In the financial market, prices of course jump, skip and leap, sharply moving up and down, i.e., there are lots of discontinuity in the movement of financial prices. In comparison, in classical physics, in a perfect gas, e.g., as molecules collide and exchange heat, their billions of individually infinitesimal transactions together produce a true "average" temperature, around which smooth gradients move up or down the scale. However, in a financial market, the news which influences an investor could be minor or major. The investor's buying power could be insignificant or market-moving. His decision could be based on an instantaneous change of decision, from bull to bear and back again. This results in a much wilder distribution of price changes - not just movements of prices but also price dislocations, which are especially noticeable in our information age with the instantaneous broadcasts by trading-room screen, television and internet. Market influencing news such as a terrorist attack or political change in a country flashes across the world to millions of investors in seconds. The investors could act on the news, not bit by bit in a progressive wave, as conventional theorists normally assume, but all at once and instantaneously. The effect could be exhilarating or heart-stopping, depending on whether one gains or loses.

A sudden price drop could cause panic in investors. The mutual fund industry sometimes takes extraordinary measures to "manage" emotions. For example, in 2000 a Milwaukee mutual fund company, Heartland Advisors Inc., hit turbulence when the market value of some of its bond investments nose-dived to $80 per $100 face value, from as high as $98, which did not show up immediately in its daily price reports and instead, according to the Securities and Exchange Commission (SEC), the fund's data supplier recorded a long, slow and gentle decline at 50 cents a day over a period of weeks. When word finally got out, Heartland investors panicked and stampeded to the exits. The data supplier was later sued by SEC but the case was settled without the data supplier admitting or denying the charges.

Discontinuity in the financial market could also bring profits, i.e., turbulence in the financial market is not always bad. The New York Stock Exchange has had, for more than a century, a system of "specialists" who are traders on the exchange floor who each specialises in the shares of a few companies, maintaining an order book, and, when the buys and the sells do not match, step in with their own money to complete trades. According to the rule, their job is to "ensure market continuity". They have however lately come into disrepute in the post-bubble scandals which have engulfed most of Wall Street. In the SEC study of a 1997 financial market collapse, it was discovered that specialists in the most tumultuous 24 minutes were powerful net buyers, with the volume of their purchases exceeding their sales by a ratio of 2.06. These were good bets for prices did recover.

It is evident that investing in the financial market is extremely risky due to discontinuities, turbulence and unpredictability in the market, wherein instead of the smooth flow of prices expected by the normal investor sudden and sharp changes in prices, i.e., discontinuities, often occur. The same could be said about turbulence in the physical world, for instance turbulence in fluids, only that in the case of the physical world it is probably worse. Take the case of weather forecasting, for example. It is not possible to forecast the weather more than a few days in advance even with very powerful computers. With just a few days of forecasting the weather still gives nasty surprises. This is due to the equations which model the weather being nonlinear, i.e., they involve the variables multiplied together, not just the variables themselves.

The theory behind the mathematics of weather forecasting was developed by Claude Navier in 1821 and George Gabriel Stokes in 1845. The Navier-Stokes equations are of very great interest to scientists, who are keen to unlock their secrets. When the Navier-Stokes equations are applied to the problem of fluid flow, they reveal much about the steady movements of the upper atmosphere. But the equations fail when applied near the earth's surface where air flow creates turbulence.

Though a lot is known about linear systems of equations, the Navier-Stokes equations contain nonlinear terms which render them intractable. The only practical way of solving the Navier-Stokes equations, which depend on initial conditions, is to do so numerically by utilising powerful computers.

Differential equations, such as the Navier-Stokes equations, could only make forecasts on phenomena characterised by smooth, regular, continuous flows, which turbulence is definitely not - turbulence, on the other hand, is characterised by great irregularities, discontinuities, disruptions and sharp jumps. With smooth, regular, continuous flows, which are each graphically represented by a smooth, continuous curve with gentle gradients, it would be possible to extrapolate and interpolate, i.e., forecasts are possible. This is not the case with turbulence, which does not display any discernable, set pattern or regularity at all. Hence, the Navier-Stokes equations fail when there is turbulence.

The solutions for turbulent flows therefore have to be left to the experimentalist and are not attempted by solving the Navier-Stokes equations.

If a scientist could produce an equation for forecasting turbulence, he could also probably apply the same equation with some modifications in forecasting the winning numbers of a lottery and he would be very wealthy and would very probably keep the equation a secret, for the uncertainties, randomness and unpredictability of turbulence and lotteries appear similar. Such a happy winner of lotteries has not happened so far, which says something about the uncertainties, randomness and unpredictability of turbulence as well.

In fact, the implicit trust in differential equations such as the famous Black-Scholes formula with its bell-curve assumption in making financial forecasts has led a number of financial organisations into trouble, e.g., the case of Long-Term Capital Management LP (LTCM), a hedge fund set up in 1993 by two Nobel laureates, Robert Merton and Myron Scholes, and some heavy-weight Wall Street bond traders, which had at one point 25 Ph.D.'s on its payroll and was possibly the best academic finance department in the world, which had made colossal losses caused by market turbulence and volatility leading to bankruptcy and had to be bailed out by several banks reluctantly through a $3.625 billion takeover at the behest of the Federal Reserve Board, which was concerned about a wave of bankruptcies if LTCM went bust.

All this goes to show how unreliable mathematics, in this case calculus, could be when there are wild swings, volatilities, discontinuities and irregularities as when there is turbulence, be it in the financial world or the physical world.

6

PRIMES, TWIN PRIMES AND SO ON

The primes, including the twin primes and the other prime pairs, are the building-blocks of the integers. Euclid's proof of the infinitude of the primes has generally been regarded as elegant. It is a proof by contradiction, or, *reductio ad absurdum*, and it relies on an algorithm which will always bring in larger and larger primes, an infinite number of them. However, the proof is also subtle and has been misinterpreted by some with one well-known mathematician even remarking that the algorithm might not work for extremely large numbers. A long unsettled related problem, the twin primes conjecture, has also aroused the interest of many researchers. Some important facts on the twin primes which would be of interest to prime number researchers, with some reasons that point to the infinitude of the twin primes, including a reasoning which is somewhat similar to Euclid's proof of the infinity of the primes are presented; very importantly, two algorithms (refer to Appendix 3) for sieving out the twin primes from the infinite list of the integers are also presented, which would be of interest to cryptographers and even computer programmers.

Introduction

In 1919, Viggo Brun (1885 - 1978) proved that the sum of the reciprocals of the twin primes converges to Brun's constant:

$$1/3 + 1/5 + 1/7 + 1/11 + 1/13 + 1/17 + 1/19 + \ldots = 1.9021605 \ldots$$

It is evident that the twin primes thin out as infinity is approached. The problem of whether there is an infinitude of twin primes is an inherently difficult one to solve, as infinity (normally symbolised by: ∞) is a difficult concept and is against common sense. It is impossible to count, calculate or live to infinity, perhaps with the exception of God. Infinity is a nebulous idea and appears to be only an abstraction devoid of any actual practical meaning. How do we quantify infinity? How big is infinity? We could either attempt to prove that the twin primes are finite, or, infinite. If the twin primes were finite, how could we prove that a particular pair of twin primes is the largest existing pair of twin primes, and, if they were infinite, how could we prove that there are always larger and larger pairs of them? It is evidently difficult to prove either, with the former appearing more difficult to prove as the odds seem against it. Some reasons in support of the latter, i.e., the infinitude of the twin primes, are provided.

Twin Primes

The following chain of reasoning points to the possibly infinity of the twin primes:-

Let 3, 5, 7, 11, 13, 17, 19, ..., n - 2, n be the list of consecutive primes, wherein n & n - 2 are assumed to be the largest existing twin primes pair, within the infinite list of the primes.

Let $3 \times 5 \times 7 \times 11 \times 13 \times 17 \times 19 = a$.

Lemma. (a x ... x n - 2 x n) - 2, &, (a x ... x n - 2 x n) - 4 will never be divisible by any of the consecutive primes in the list: 3, 5, 7, 11, 13, 17, 19, ..., n - 2, n, whether they are prime or composite. (See Appendix 1 for the proof.)

This implies that:

If (a x ... x n - 2 x n) - 2 &/V (a x ... x n - 2 x n) - 4 are prime, then:

(a x ... x n - 2 x n) - 2 > (a x ... x n - 2 x n) - 4 > n > n - 2

If (a x ... x n - 2 x n) - 2 &/V (a x ... x n - 2 x n) - 4 are non-prime/composite, then:

(a) each prime factor, e.g., y below, of (a x ... x n - 2 x n) - 2 > n > n - 2
(b) each prime factor, e.g., z below, of (a x ... x n - 2 x n) - 4 > n > n - 2

(a x ... x n - 2 x n) - 2 = prime V composite   (1)

(a x ... x n - 2 x n) - 4 = prime V composite   (2)

(1) & (2) = twin primes, if both (1) & (2) are prime

(1) & (2) > n & n - 2

Let Y represent the prime factors of (a x ... x n - 2 x n) - 2 if (a x ... x n - 2 x n) - 2 is not prime (i.e., it is composite), each prime factor may pair up with another prime which differs from it by 2 to form twin primes. Let y = prime factor in Y.

y & y +/- 2 = twin primes, if y +/- 2 is prime

y & y +/- 2 > n & n - 2

Let Z represent the prime factors of (a x ... x n - 2 x n) - 4 if (a x ... x n - 2 x n) - 4 is not prime (i.e., it is composite), each prime factor may pair up with another prime which differs from it by 2 to form twin primes. Let z = prime factor in Z.

z & z +/- 2 = twin primes, if z +/- 2 is prime

z & z +/- 2 > n & n - 2

Therefore: (a x ... x n - 2 x n) - 2 > (a x ... x n - 2 x n) - 4 > y V y +/- 2 V z V z +/- 2 > n > n - 2

By the above, the following, which implies that n & n - 2 are the largest existing twin primes pair, is an oddity:

n > n - 2 > (a x ... x n - 2 x n) - 2 > (a x ... x n - 2 x n) - 4 > y V y +/- 2 V z V z +/- 2

It seems that no n & n - 2 in any list of consecutive primes can ever possibly be the largest existing twin primes pair and larger twin primes than them can always be found by applying the same mathematical logic (as is described in Appendix 1), e.g., by utilising the evidently effective Algorithm 1, or, Algorithm 2 described in Appendix 3. That is, a largest existing twin primes pair is an oddity, which suggests that the twin primes are infinite. It is possible to find larger twin primes than n & n - 2 no matter how large n & n - 2 are, with the following formulae involving the list of consecutive primes: (a x ... x n) - 2 & (a x ... x n) - 4, which by the nature of their composition are capable of generating new primes/twin primes which will always be larger than n & n - 2 (see Appendix 1); this operation is part of Algorithm 1 described in Appendix 3. This is an indirect argument or argument by contradiction (*reductio ad absurdum*) for the infinity of the twin primes, for our assumption of n & n - 2 as the largest existing twin primes pair will be contradicted by the discovery of larger twin primes, implying the infinity of the twin primes. Again, by applying the same mathematical logic (described in Appendix 1), by way of this evidently effective Algorithm 1 in Appendix 3, and going one step further, we can find that many twin odd integers found between n and (a x ... x n) - 2, which differ from one another by 2 and are not divisible by any of the primes in the list of consecutive primes: 3, 5, 7, 11, 13, 17, 19, ..., n, will be twin primes larger than n & n - 2, our assumed largest existing twin primes pair, which are further or more contradictions of this assumption. In this manner, i.e., by resorting to Algorithm 1 in Appendix 3, by continually adding more and more consecutive primes to the list of consecutive primes: 3, 5, 7, 11, 13, 17, 19, ..., n, i.e., continually extending the value of n, and utilising the formula: (a x ... x n) - 2, as well as the formula: (a x ... x n) - 4, to perform the computations a la Algorithm 1 in Appendix 3, many larger and larger twin primes can be found, all the way to infinity, in parallel with the infinitude of the list of consecutive primes: 3, 5, 7, 11, 13, 17, 19, ..., of which the twin primes are a part together with other primes pairs, wherein the twin primes are not at all likely to be finite (as is evident from Appendix 2) and can be expected to be infinite. (Algorithm 2 in Appendix 3 may also be utilised for this purpose but it is evidently a longer and less efficient method. It is important to note that Algorithm 1, as well as Algorithm 2, would be able to fish out or generate a big quantity of twin primes which are larger than n & n - 2, our assumed largest existing twin primes pair, giving rise to a large number of contradictions of this assumption, thus much further implying the infinitude of the twin primes, as is shown in great detail in Appendix 3, but is only briefly described above.) A largest existing twin primes pair would seem an impossibility. It could be concluded that the twin primes are infinite.

Conclusion

The list of the twin primes appears infinite, which is ratified by the strong anecdotal evidence provided in Appendix 2. Very importantly, there are two algorithms provided in Appendix 3 for fishing out the twin primes from the infinite list of the integers; a strong argument supporting the infinity of the twin primes without any means or algorithms for finding them would be less strong than an argument backed by some algorithms for finding the twin primes - such an argument strongly backed by the algorithms would be a constructive argument. These two algorithms would be able to sieve out many, many twin primes from the list of the integers, to infinity. Finally, it should be noted that as "closest next-door neighbours" among the primes, the infinity of the twin primes seems evident, a point which is explained in Appendix 4.

Remarks

There are 376 pairs of twin primes (752 primes) found within the 2,500 consecutive primes from 2 to 22,307 - this means that 30.08%, which is sizeable, of the 2,500, not a small quantity, consecutive primes are twin primes. 3, 5 & 7 are the only "triple" primes found. There is no regularity in pattern in the appearance of the twin primes, except that the intervals between consecutive twin primes vary greatly by from 4 integers to 370 integers - the intervals between the consecutive twin primes increase and decrease, and, then increase and decrease again, by turns, giving rise to a graph that is characterised by many peaks, i.e., the curve is rough and nonlinear, making its description (hence, forecast of the twin primes) by differential equations practically impossible.

  The argument used here to support the twin primes' infinity is the indirect (*reductio ad absurdum*) method, which had been used by Euclid and other mathematicians after him. Logically, one or two examples of "contradiction" should be sufficient evidence of infinity, for it does not make sense to have a need for an infinite number of cases of "contradiction", as our argument would then have to be infinitely and impossibly long, an absurdity. This method of argument is "argument by implication" as a result of "contradiction" - which is a "short-cut" and smart way in indicating infinity, instead of "showing infinity by counting to infinity", which is ludicrous, and, impossible. Hence, one or two cases of "contradiction" should be sufficient for implying that there would be an infinitude of twin primes, which of course also tacitly implies that there would be an infinitude of the number of cases of such "contradiction". (Euclid evidently had this logical point in mind when he formulated the indirect (*reductio ad absurdum*) argument for the infinity of the primes.) This method of argument had been cleverly used by a number of mathematicians, not the least by the great German mathematician, David Hilbert. For example, Hilbert had used an indirect method (the "*reductio ad absurdum*" argument) to support Gordan's Theorem without having to show an actual "construction", an argument which had been accepted by his peers.

Two algorithms for generating or sieving many of the twin primes in any range of odd numbers are presented - by utilising any of these two algorithms (preferably the evidently more efficient Algorithm 1), we will be able to find many twin primes which are all larger than those in any chosen list of consecutive primes, i.e., we will be able to generate many larger and larger twin primes. This is indeed significant. There is evidently some deep meaning in the ease with which the twin primes turn up, as is shown here. It is thus evident that the twin primes are an inherent characteristic of the infinite prime numbers (as well as odd numbers), a characteristic which could be regarded as "self-similar" or "fractal". A twin primes pair is in effect any pair of odd numbers which differ from one another by 2 and are indivisible by any number except itself, the negative of itself, +1 and -1 (i.e., the pair of odd numbers are prime numbers). Any consecutive odd numbers or odd numbers that differ from one another by 2 are therefore potential prime numbers, as well as potential twin primes, and, the likelihood of them being prime is infinite (vide Euclid's proof and Dirichlet's Theorem), i.e., the primes will always be found amongst them and will be there all the way to infinity (the primes being evidently the "atoms" or building-blocks of all the whole numbers or integers, i.e., all the odd numbers and even numbers - every odd number or integer is either a prime number or composite of prime numbers (i.e., the integer has prime factors), and, every even number is the sum of two prime numbers (vide the Goldbach conjecture which, it appears, practically all mathematicians believe to be true), as well as the product of prime numbers (composite)); hence, the likelihood of them being twin primes is infinite as well (the twin primes being an inherent property of the infinite prime numbers - as well as odd numbers - the twin primes can in fact be likened to next-door neighbours, which are a common, expected thing (see Appendix 4)).

So far, there has not been any indication or confirmation that the number of twin primes is finite and the so-called largest existing pair of twin primes has not been found and confirmed (which of course would be impossible to find and confirm if the twin primes were infinite). On the other hand, practically everyone could intuit that the number of twin primes is infinite.

Due to the evident effectiveness of the two algorithms described in Appendix 3 in bringing in larger and larger twin primes, the above argument for the infinitude of the twin primes is not only an indirect argument or argument by contradiction (*reductio ad absurdum*), importantly, it is also a constructive argument. It should be noted that the characteristic of a mountain or infinite volume of sand is reflected in the characteristic of some grains of sand found there so that studying the characteristic of some grains of sand found there is enough for deducing the characteristic of the mountain or infinite volume of sand, to ascertain the quality of a batch of products it is only necessary to inspect some carefully selected samples from that batch of products and not every one of the products and to carry out a population census, i.e., find out the characteristics of a population, it is only necessary to carry out a survey on some carefully selected respondents and not the whole population. With any of these two algorithms (preferably the evidently more efficient Algorithm 1), in like manner, by the same principle, we could carry out a study of a carefully selected list of integers and their associated primes and twin primes and deduce by induction whether the twin primes would always turn up, appear infinitely, in the list which is itself infinite - this act is similar to extrapolation.

Appendix 1

Note: The (only) even prime 2 is omitted from the list of consecutive primes: 3, 5, 7, 11, 13, 17, 19, ..., n - 2, n stated herein, wherein n & n - 2 are assumed to be the largest existing twin primes pair.

The list of newly created primes, and, twin primes for n = 5, 7, 11, 13, 17, 19, ... (n = 19 being the maximum limit achievable with a hand-held calculator) is as follows:-

1] For n = 5, we get the following new primes/new twin primes:

   (3 x 5) - 2 = 13   (α)
   (3 x 5) - 4 = 11   (β)

2] For n = 7, we get the following new primes/new twin primes:

   (3 x 5 x 7) - 2 = 103   (α)
   (3 x 5 x 7) - 4 = 101   (β)

3] For n = 11, we get the following new primes/new twin primes:

   (3 x 5 x 7 x 11) - 2 = 1,153   (α)
   (3 x 5 x 7 x 11) - 4 = 1,151   (β)

4] For n = 13, we get the following new prime and composite number with its prime factors:

   (3 x 5 x 7 x 11 x 13) - 2 = 15,013   (α) - Prime Number
   (3 x 5 x 7 x 11 x 13) - 4 = 15,011   (β) - Composite Number (= 17 x 883, with 17 pairing with 19 to form a twin primes pair and 883 pairing with 881 to form another twin primes pair)

5] For n = 17, we get the following new primes/new twin primes:

   (3 x 5 x 7 x 11 x 13 x 17) - 2 = 255,253   (α)
   (3 x 5 x 7 x 11 x 13 x 17) - 4 = 255,251   (β)

6] For n = 19, we get the following new prime and composite number with its prime factors:

   (3 x 5 x 7 x 11 x 13 x 17 x 19) - 2 = 4,849,843   (α) - Prime Number
   (3 x 5 x 7 x 11 x 13 x 17 x 19) - 4 = 4,849,841   (β) - Composite Number (= 43 x 112,787, with 43 pairing with 41 to form a twin primes pair while 112,787 is a stand-alone prime)

.
.
.

Results Of α And β Above

1) α above generates 6 new primes (13; 103; 1,153; 15,013; 255,253; 4,849,843), nil composite numbers.
2) β above generates 4 new primes (11; 101; 1,151; 255,251), 2 composite numbers (15,011 = 17 x 883; 4,849,841 = 43 x 112,787).
3) α and β above together produce 4 pairs of new twin primes (13 & 11; 103 & 101; 1,153 & 1,151; 255,253 & 255,251).
4) The prime factors of α and β above form 3 pairs of new twin primes with prime partners which differ from them by 2 (19 & 17; 43 & 41; 883 & 881).
5) All the new twin primes in (3) and (4) above are larger than n & n - 2, the assumed largest existing twin primes pair, which is an indirect argument for the infinitude of the twin primes.

Why It Is Impossible For Any n & n - 2 To Be The Largest Existing Twin Primes Pair

α = (3 x 5 x 7 x 11 x 13 x 17 x 19 x ... x n) - 2, and, β = (3 x 5 x 7 x 11 x 13 x 17 x 19 x ... x n) - 4 will never be divisible by any of the consecutive prime numbers in the list: 3, 5, 7, 11, 13, 17, 19, ..., n, whether they are prime or composite (non-prime and divisible by prime numbers or prime factors). This means that none of the consecutive prime numbers in the list: 3, 5, 7, 11, 13, 17, 19, ..., n can ever be factors of α and β, and, α and β must be new primes/twin primes larger than all the consecutive prime numbers in the list: 3, 5, 7, 11, 13, 17, 19, ..., n, or, if they were composite (non-prime and divisible by prime numbers or prime factors), their prime factors (and "twin prime" partners which differ from them by 2) must be larger than all the consecutive prime numbers in the list: 3, 5, 7, 11, 13, 17, 19, ..., n. *This is a very important mathematical logic, which needs to be grasped in order to understand the argument.*

This all implies that no n & n - 2 (if n - 2 were also a prime number) in any list of consecutive prime numbers can ever possibly be the largest existing twin primes pair, since all the new primes/twin primes produced or generated by α and β will always be larger than n & n - 2. That is, a largest existing twin primes pair is an impossibility, which implies the infinitude of the list of the primes/twin primes.

In other words, by the mathematical logic stated above, which explains why all the new primes/twin primes, which α and β by the nature of their composition are capable of producing or generating, will always be larger than n & n - 2, no n & n - 2 in any list of consecutive prime numbers: 3, 5, 7, 11, 13, 17, 19, ..., n can ever possibly be the largest existing twin primes pair, i.e., a largest existing twin primes pair is an impossibility, thus implying the infinitude of the list of the twin primes. This is a very important inference.

Regardless of how long the list of the twin primes pairs is, it is possible to find some new twin primes pairs which will always be larger than n & n - 2, our assumed largest existing twin primes pair - the largest twin primes pair in our assumed finite list of the twin primes pairs, with α and β, which is an indirect argument for the infinity of the twin primes. In fact, by the same principle, many twin odd integers found between n and α, which differ from one another by 2 and are not divisible by any of the primes in the list of consecutive primes: 3, 5, 7, 11, 13, 17, 19, ..., n, will be twin primes pairs larger than n & n - 2, our assumed largest existing twin primes pair, which is a contradiction of this assumption, hence implying the infinitude of the twin primes. (Refer to Algorithm 1, as well as Algorithm 2, in Appendix 3.)

Appendix 2

Anecdotal Evidence Of The Infinity Of The Twin Primes

TOP TWIN PRIMES IN 2000, 2001, 2007 & 2009
In the year 2000, $4648619711505 \times 2^{60000} \pm 1$ (18,075 digits) had been the top twin primes pair which had been discovered. In the year 2001, it only ranked eighth in the list of top 20 twin primes pairs, with $318032361 \cdot 2^{107001} \pm 1$ (32,220 digits) topping the list. In the year 2007, in the list of top 20 twin primes pairs, $318032361 \cdot 2^{107001} \pm 1$ (32,220 digits) ranked eighth, while $4648619711505 \times 2^{60000} \pm 1$ (18,075 digits) was nowhere to be seen; $2003663613*2^{195000}-1$ and $2003663613*2^{195000}+1$ (58,711 digits), which was discovered on January 15, 2007, by Eric Vautier (from France) of the Twin Prime Search (TPS) project in collaboration with PrimeGrid (BOINC platform), was at the top of the list. As at August 2009, $65516468355 \cdot 2^{333333}-1$ and $65516468355 \cdot 2^{333333}+1$ (100,355 digits) is at the top of the list of top 20 twin primes pairs, while $318032361 \cdot 2^{107001} \pm 1$ (32,220 digits) ranks 11th, and, $2003663613*2^{195000}-1$ and $2003663613*2^{195000}+1$ (58,711 digits) ranks second in this list.

We can expect larger twin primes than these extremely large twin primes, much larger ones, infinitely larger ones, to be discovered in due course.

LIST OF PRIMES PAIRS FOR THE FIRST 2,500 CONSECUTIVE PRIMES, 2 TO 22,307, RANKED ACCORDING TO THEIR FREQUENCIES OF APPEARANCE

S. No.	Ranking	Prime Pairs	No. Of Pairs	Percentage
(1)	1	primes pair separated by 6 integers	482	19.29 %
(2)	2	primes pair separated by 4 integers	378	15.13 %
(3)	3	primes pair separated by 2 integers (t. p.)	376	15.05 %
(4)	4	primes pair separated by 12 integers	267	10.68 %
(5)	5	primes pair separated by 10 integers	255	10.20 %
(6)	6	primes pair separated by 8 integers	229	9.16 %
(7)	7	primes pair separated by 14 integers	138	5.52 %
(8)	8	primes pair separated by 18 integers	111	4.44 %
(9)	9	primes pair separated by 16 integers	80	3.20 %
(10)	10	primes pair separated by 20 integers	47	1.88 %
(11)	11	primes pair separated by 22 integers	46	1.84 %
(12)	12	primes pair separated by 30 integers	24	0.96 %
(13)	13	primes pair separated by 28 integers	19	0.76 %
(14)	14	primes pair separated by 24 integers	16	0.64 %
(15)	15	primes pair separated by 26 integers	10	0.40 %
(16)	16	primes pair separated by 34 integers	9	0.36 %
(17)	17	primes pair separated by 36 integers	5	0.20 %
(18)	18	primes pair separated by 32 integers	2	0.08 %
(19)	18	primes pair separated by 40 integers	2	0.08 %
(20)	19	primes pair separated by 42 integers	1	0.04 %
(21)	19	primes pair separated by 52 integers	1	0.04 %

Total No. Of Primes Pairs In List: 2,498

It is evident in the above list that the primes pairs separated by 6 integers, 4 integers and 2 integers (twin primes), among the 21 classifications of primes pairs separated by from 2 integers to 52 integers (primes pairs separated by 38 integers, 44 integers, 46 integers, 48 integers & 50 integers are not among them, but, they are expected to appear further down in the infinite list of the primes), are the most dominant, important. There is a long list of other primes pairs, besides those shown in the above list, which also play a part as the building-blocks of the infinite list of the integers.

The list of the integers is infinite. The list of the primes is also infinite. The infinite primes are the building-blocks of the infinite integers - the infinite odd integers are all either primes or composites of primes, and, the infinite even integers, except for 2 which is a prime, are all also composites of primes. Therefore, all the primes pairs separated by the integers of various magnitudes, as described above, can never all be finite. If there is any possibility at all for any of these primes pairs to be finite, there is only the possibility that a number of these primes pairs are finite (but never all of them). However, will it have to be the primes pairs separated by 2 integers or twin primes (which are the subject of our investigation here), which are the only primes pair, or, one among a number of primes pairs, which are finite? Why question only the infinity of the primes pairs separated by 2 integers, the twin primes? Are not the infinities of the primes pairs separated by 8 integers and more, whose frequencies of appearance are lower, as compared to those of the primes pairs which are separated by 6, 4 and 2 integers respectively, in the above list of primes pairs, more questionable? Why single out only the twin primes? (There are at least 18 other primes pairs, separated by from 8 integers to 52 integers, whose respective infinities should be more suspect, as is evident from the above list of primes pairs, if any infinities should be doubted. Evidently, the primes pairs separated by 2 integers (twin primes) are not at all likely to be finite.)

The above represents anecdotal evidence that the twin primes are infinite, which is a ratification of the reasoning given earlier.

Appendix 3

The following algorithms will be able to generate or sieve all the twin primes in any range of odd numbers which are all larger than those in the list of known consecutive primes/twin primes; these 2 important algorithms will provide plenty of numerical evidence that the twin primes are infinite:-

Algorithm 1

We would provide an example with Items (1) to (3) from the following list of products of consecutive primes/twin primes, which should be sufficient for our purpose here:-

1) $3 \times 5 = 15$
2) $3 \times 5 \times 7 = 105$
3) $3 \times 5 \times 7 \times 11 = 1,155$
4) $3 \times 5 \times 7 \times 11 \times 13 = 15,015$
5) $3 \times 5 \times 7 \times 11 \times 13 \times 17 = 255,255$
6) $3 \times 5 \times 7 \times 11 \times 13 \times 17 \times 19 = 4,849,845$
   .
   .
   .

The example is as follows:-

1) For $3 \times 5 = 15$, we would find all the consecutive pairs of odd numbers between 5 & 15 which differ from one another by 2 and are not divisible by any of the consecutive primes/twin primes 3 & 5 in the list of consecutive primes/twin primes 3×5 whose product is 15.

   There is only 1 pair of odd numbers between 5 & 15 which differ from one another by

2 and are not divisible by the consecutive primes/twin primes 3 & 5 in the list of consecutive primes/twin primes 3 x 5 - they are the twin primes 11 & 13.

2) Similarly, for 3 x 5 x 7 = 105, we would find all the consecutive pairs of odd numbers between 7 & 105 which differ from one another by 2 and are not divisible by any of the consecutive primes/twin primes 3, 5 & 7 in the list of consecutive primes/twin primes 3 x 5 x 7 whose product is 105.

The consecutive pairs of odd numbers between 7 & 105 which differ from one another by 2 and are not divisible by the consecutive primes/twin primes 3, 5 & 7 are the following consecutive twin primes:

(a) 11 & 13
(b) 17 & 19
(c) 29 & 31
(d) 41 & 43
(e) 59 & 61
(f) 71 & 73
(g) 101 & 103

3) Similarly, in this final case, for 3 x 5 x 7 x 11 = 1,155, we would find all the consecutive pairs of odd numbers between 11 & 1,155 which differ from one another by 2 and are not divisible by any of the consecutive primes/twin primes 3, 5, 7 & 11 in the list of consecutive primes/twin primes 3 x 5 x 7 x 11 whose product is 1,155.

Many of the consecutive pairs of odd numbers between 11 & 1,155 which differ from one another by 2 and are not divisible by the consecutive primes/twin primes 3, 5, 7 & 11 are twin primes (while the rest are primes larger than 3, 5, 7 & 11 and/or composite numbers whose prime factors are each larger than 3, 5, 7 & 11), some of which are as follows:

(a) 17 & 19
(b) 29 & 31
(c) 41 & 43
(d) 59 & 61
(e) 71 & 73
(f) 101 & 103
(g) 107 & 109
(h) 137 & 139
(i) 149 & 151
(j) 179 & 181
(k) Etc. to 1,151 & 1,153

In this way, we would also be able to achieve the following:-

1) For 3 x 5 x 7 x 11 x 13 = 15,015, find all the consecutive twin primes between 13 and 15,015.
2) For 3 x 5 x 7 x 11 x 13 x 17 = 255,255, find all the consecutive twin primes between 17 and 255,255.
3) For 3 x 5 x 7 x 11 x 13 x 17 x 19 = 4,849,845, find all the consecutive twin primes between 19 and 4,849,845.

⋮

Algorithm 2
We would, similar to Algorithm 1 above, also provide an example with Items (1) to (3) from the following list of products of consecutive primes/twin primes, which should be sufficient for our purpose here:-

1) 3 x 5 = 15
2) 3 x 5 x 7 = 105
3) 3 x 5 x 7 x 11 = 1,155
4) 3 x 5 x 7 x 11 x 13 = 15, 015
5) 3 x 5 x 7 x 11 x 13 x 17 = 255,255
6) 3 x 5 x 7 x 11 x 13 x 17 x 19 = 4,849,845

⋮

The example is as follows:-

1) For 3 x 5 = 15, we would first find all the consecutive pairs of even numbers between 5 & 15 which differ from one another by 2 and are not divisible by any of the consecutive primes/twin primes 3 & 5 in the list of consecutive primes/twin primes 3 x 5. Then we deduct each of these consecutive pairs of even numbers which are not divisible by any of the consecutive primes/twin primes 3 & 5 from the product of these consecutive primes/twin primes 3 x 5 which is 15. The results would each be 1 pair of twin primes (which are each larger than 3 & 5), 1 prime (which is larger than 3 & 5) & 1 composite of primes (whose prime factors are each larger than 3 & 5), or, 2 composites of primes (whose prime factors are each larger than 3 & 5). In this way, we would be able to find all the consecutive twin primes between 5 & 15.

There are no pairs of even numbers between 5 & 15 which differ from one another by 2 and are not divisible by any of the consecutive primes/twin primes 3 & 5 in the list of consecutive primes/twin primes 3 x 5 - the exception is the smallest pair of even numbers 2 & 4, which are not divisible by any of the consecutive primes/twin primes 3 & 5 in the list of consecutive primes/twin primes 3 x 5.

The following is the result after we deduct this pair of even numbers 2 & 4 which are not divisible by any of the consecutive primes/twin primes 3 & 5 from the product of these consecutive primes/twin primes 3 x 5 which is 15:

(a) 15 - 2 & 15 - 4: 13 & 11 (twin primes)

2) Similarly, for 3 x 5 x 7 = 105, we would first find all the consecutive pairs of even numbers between 7 & 105 which differ from one another by 2 and are not divisible by any of the consecutive primes/twin primes 3, 5 & 7 in the list of consecutive primes/twin primes 3 x 5 x 7, which are as follows (the smallest pair of even numbers 2 & 4 is the exception and they are not divisible by any of the consecutive primes/twin primes 3, 5 & 7 in the list of consecutive primes/twin primes 3 x 5 x 7):

(a) 2 & 4
(b) 32 & 34
(c) 44 & 46
(d) 62 & 64
(e) 74 & 76
(f) 86 & 88
(g) 92 & 94

Then we deduct each of these consecutive pairs of even numbers which are not divisible by any of the consecutive primes/twin primes 3, 5 & 7 from the product of these consecutive primes/twin primes 3 x 5 x 7 which is 105. The results would each be 1 pair of twin primes (which are each larger than 3, 5 & 7), 1 prime (which is larger than 3, 5 & 7) & 1 composite of primes (whose prime factors are each larger than 3, 5 & 7), or, 2 composites of primes (whose prime factors are each larger than 3, 5 & 7). In this way, we would be able to find all the consecutive twin primes between 7 & 105, which are as follows:

(a) 105 - 2 & 105 - 4:   103 & 101 (twin primes)
(b) 105 - 32 & 105 - 34:   73 & 71 (twin primes)
(c) 105 - 44 & 105 - 46:   61 & 59 (twin primes)
(d) 105 - 62 & 105 - 64:   43 & 41 (twin primes)
(e) 105 - 74 & 105 - 76:   31 & 29 (twin primes)
(f) 105 - 86 & 105 - 88:   19 & 17 (twin primes)
(g) 105 - 92 & 105 - 94:   13 & 11 (twin primes)

3) Similarly, in this final case, for 3 x 5 x 7 x 11 = 1,155, we would first find all the consecutive pairs of even numbers between 11 & 1,155 which differ from one another by 2 and are not divisible by any of the consecutive primes/twin primes 3, 5, 7 & 11 in the list of consecutive primes/twin primes 3 x 5 x 7 x 11, some of which are as follows
(the smallest pair of even numbers 2 & 4 is the exception and they are not divisible by any of the consecutive primes/twin primes 3, 5, 7 & 11 in the list of consecutive primes/twin primes 3 x 5 x 7 x 11):

(a) 2 & 4
(b) 32 & 34
(c) 62 & 64
(d) 74 & 76
(e) 92 & 94
(f) 116 & 118
(g) 122 & 124
(h) 134 & 136
(i) Etc. to 1,136 & 1,138

Next we deduct each of these consecutive pairs of even numbers which are not divisible by any of the consecutive primes/twin primes 3, 5, 7 & 11 from the product of these consecutive primes/twin primes 3 x 5 x 7 x 11 which is 1,155. The results would each be 1 pair of twin primes (which are each larger than 3, 5, 7 & 11), 1 prime (which is larger than 3, 5, 7 & 11) & 1 composite of primes (whose prime factors are each larger than 3, 5, 7 & 11), or, 2 composites of primes (whose prime factors are each larger than 3, 5, 7 & 11). In this way, we would be able to find all the consecutive twin primes between 11 & 1,155, some of which are as follows:

(a) 1,155 - 2 & 1,155 - 4:   1,153 & 1,151 (twin primes)
(b) 1,155 - 32 & 1,155 - 34:   1,123 (prime) & 1,121 (composite of primes which
    are each larger than 3, 5, 7 &
    11 = 19 x 59)
(c) 1,155 - 62 & 1,155 - 64:   1,093 & 1,091 (twin primes)
(d) 1,155 - 74 & 1,155 - 76:   1,081 & 1,079
    (composite of primes   (composite of
    which are each larger   primes which are
    than 3, 5, 7 & 11 =   each larger than
    23 x 47)   3, 5, 7 & 11 =
    13 x 83)
(e) 1,155 - 92 & 1,155 - 94:   1,063 & 1,061 (twin primes)
(f) 1,155 - 116 & 1,155 - 118: 1,039 (prime) & 1,037 (composite of primes which
    are each larger than 3, 5, 7 &
    11 = 17 x 61)
(g) 1,155 - 122 & 1,155 - 124: 1,033 & 1,031 (twin primes)
(h) 1,155 - 134 & 1,155 - 136: 1,021 & 1,019 (twin primes)
(i) Etc. to 1,155 - 1,136 & 1,155 - 1,138: 19 & 17 (twin primes)

In like manner, we would also be able to achieve the following:-

1) For 3 x 5 x 7 x 11 x 13 = 15,015, find all the consecutive twin primes between 13 and 15,015.
2) For 3 x 5 x 7 x 11 x 13 x 17 = 255,255, find all the consecutive twin primes between 17 and 255,255.
3) For 3 x 5 x 7 x 11 x 13 x 17 x 19 = 4,849,845, find all the consecutive twin primes between 19 and 4,849,845.

.
.
.

By utilising any of the above algorithms (preferably the evidently more efficient Algorithm 1), we will be able to find many twin primes which are all larger than those in any chosen list of consecutive primes/twin primes, i.e., we will be able to generate many larger and larger twin primes with these algorithms.

It would evidently be difficult to accept an argument supporting the twin primes conjecture without having to confirm or check the validity of the logic by computing a sufficiently long list of twin primes, even to the extent of looking out for counter-examples. Hence, the great importance of the above algorithms.

Appendix 4

Further Remarks On The Twin Primes
We note a very important intrinsic characteristic of the primes. Like all the houses in a neighbourhood or location which are separated from each other by the number of houses between them, the primes are also separated from each other by the number of integers separating them. The closest will of course be the prime neighbours separated by 2 integers (i.e., twin primes), followed next in proximity by the prime neighbours separated by 4 integers, then by the prime neighbours separated by 6 integers, the prime neighbours separated by 8 integers, the prime neighbours separated by 10 integers, the prime neighbours separated by 12 integers, and so on, by larger and larger intervals, as is shown in Appendix 2. The twin primes are actually comparable to 2 closest neighbours living just next door to one another. There will

always be 2 closest next-door neighbours, neighbours living 2 doors away, neighbours living 3 doors away, neighbours living 4 doors away, neighbours living 5 doors away, neighbours living 6 doors away, and so on, by greater and greater intervals, in any neighbourhood or residential area; there will always be different intervals separating all the houses in a neighbourhood or location. Similarly, in the infinite list of the primes, there will always be different intervals separating all the primes, ranging from the smallest interval of 2 integers (in the case of the twin primes), 4 integers, 6 integers, 8 integers, 10 integers, 12 integers, and more and more integers, etc., which is an intrinsic characteristic of the primes. In other words, there will always be intervals of various magnitudes or sizes (i.e., intervals of various numbers of integers) between, separating, all the primes in the infinite list of the primes, and, each of these intervals of various magnitudes or sizes can be expected to be infinite as the list of the primes is infinite. The twin primes, which we are examining here, are not at all likely to be finite (as is evident from Appendix 2), and should be infinite; in fact, to say that the twin primes are finite is like saying that next-door neighbours who are closest are rare and limited, which is, in the author's opinion, absurd.

7

INTERESTING TWIN PRIMES

Many believe that the twin primes are infinite. In fact, twin primes pairs could easily be found among the integers. There is evidently no region of the natural number system so remote that it lies beyond the largest twin primes pair. It is even possible to forecast the approximate number of twin primes pairs found in any region of the natural number system.

The occurrence of twin primes pairs is evidently unpredictable or random. This means that the chance of 2 numbers x and x + 2 being prime (twin primes) is somewhat similar to the chance of getting heads on 2 successive tosses of a coin. If 2 successive tosses of a coin are independent, the chance of success of obtaining heads for the 2 successive tosses of the coin is the product of the chances of success of obtaining a head for each toss of the coin. As each coin has probability ½ of coming up heads with a toss, 2 coins would have probability ½ x ½ = ¼ of coming up a pair of heads with a toss.

The prime number theorem, which had been proven, states that if n is a large number, and we select a number x at random between 0 and n, the chance that x is prime would be approximately $1/\log n$, the larger n is, the better would be the approximation given by $1/\log n$ to the proportion of primes in the numbers up to n. Like 2 coins coming up heads, the chance that both x and x + 2 are prime (twin primes) would be approximately $1/(\log n)^2$. That is, there would be approximately $n/(\log n)^2$ twin primes pairs between 0 and n. As n goes to infinity, this fraction approaches infinity. This represents a quantitative version of the twin primes conjecture.

As x + 2 being prime depends on the fact that x is already prime, we should modify the estimate $n/(\log n)^2$ to $(1.32032..)n/(\log n)^2$.

The following is a comparison between the twin primes predicted by the above formula and the twin primes found, where the agreement is evidently very good:-

INTERVAL	TWIN PRIMES	
	PREDICTED	FOUND
100,000,000 - 100,150,000	584	601
1,000,000,000 - 1,000,150,000	461	466
10,000,000,000 - 10,000,150,000	374	389
100,000,000,000 - 100,000,150,000	309	276
1,000,000,000,000 - 1,000,000,150,000	259	276

10,000,000,000,000 - 10,000,000,150,000	221	208
100,000,000,000,000 - 100,000,000,150,000	191	186
1,000,000,000,000,000 - 1,000,000,000,150,000	166	161

All this represents numerical evidence that the twin primes are infinite as we could find more twin primes pairs whenever we look for them. But the proof is lacking.

We explain why the list of the twin primes pairs should be infinite.

Lemma 1. According to the precepts of fractal geometry and group theory, symmetry is a very important, intrinsic part of nature. There is symmetry all around us and within us. There is evident symmetry in human bodies, the structures of viruses and bacteria, polymers and ceramic materials, the permutations of numbers, the universe and many others, even the movements of prices in financial markets, the growths of populations, the sound of music, the flow of blood through our circulatory system, the behavior of people en masse, etc. In other words, regularity, pattern, order, uniformity or symmetry is evident everywhere.

The reasoning here makes use of a very important idea in fractal geometry and group theory, namely, symmetry.

A prime number is an integer which is divisible only by 1 and itself, e.g., 2, 3, 7, 19, etc. A twin primes pair are 2 primes which differ from one another by 2, e.g., 5 & 7, 11 & 13, 17 & 19, and, 29 & 31, etc. A composite number or non-prime is a product of primes or prime factors, e.g., the composite numbers 15 is the product of 2 primes, 3 and 5 (15 = 3 x 5), and 231 is the product of 3 primes, 3, 7 and 11 (231 = 3 x 7 x 11), etc. The integers or whole numbers are either primes or composites and are infinite.

The primes, which Euclid had proven to be infinite, are the atoms or building-blocks of the infinite integers or whole numbers, which comprise of the infinite list of the odd numbers that are all either primes or products of primes (i.e., composites), and, the infinite list of the even numbers that are all products of primes (i.e., composites, with the exception of 2 which is a prime, e.g., 6 = 2 x 3, 8 = 2 x 2 x 2 and 10 = 2 x 5, etc.). The infinite list of the integers or whole numbers may be classified as an infinite group, with various symmetries, subgroups and infinite elements, hidden within it. These various symmetries, subgroups and infinite elements, within this infinite group may be classified as follows:-

(1) <u>Subgroup A</u>: Infinite consecutive primes such as 2, 3, 5, 7, 11, 13, 17, 19, 23, 29, 31, etc. to infinity ...., separated by 2 integers (twin primes), 4 integers, 6 integers, 8 integers, 10 integers, etc. to infinity ...., which, incidentally, except for 2, are all odd numbers; this splitting up of the subgroup into infinite elements is shown below:

  (i) <u>Element A1</u>: Infinite list of all the primes pairs separated by 2 integers (twin primes) .... (Example: 17 & 19)
  (ii) <u>Element A2</u>: Infinite list of all the primes pairs separated by 4 integers/1 odd composite - single composite .... (Example: 79 & 83 separated by 81)
  (iii) <u>Element A3</u>: Infinite list of all the primes pairs separated by 6 integers/2 consecutive odd composites - twin composites .... (Example: 47 & 53 separated by 49 & 51)
  (iv) <u>Element A4</u>: Infinite list of all the primes pairs separated by 8 integers/3 consecutive odd composites - "triple" composites .... (Example: 359 & 367 separated by 361, 363 & 365)

(v) <u>Element A5</u>: Infinite list of all the primes pairs separated by 10 integers/4 consecutive odd composites - "four-ple" composites ………. (Example: 709 & 719 separated by 711, 713, 715 & 717)

.
.
.

(2) <u>Subgroup B</u>: Infinite consecutive odd composites such as 9, 15, 21, 25, 27, 33, 35, 39, 45, 49, etc. to infinity ………, of "infinite sizes" sandwiched between 2 primes; this splitting up of the subgroup into infinite elements is shown below:

(i) <u>Element B1</u>: Infinite list of all "1 odd composite sandwiched between 2 primes - single composite" ………. (Example: 9 sandwiched between the primes 7 & 11)
(ii) <u>Element B2</u>: Infinite list of all "2 consecutive odd composites sandwiched between 2 primes - twin composites" ………. (Example: 253 & 255 sandwiched between the primes 251 & 257)
(iii) <u>Element B3</u>: Infinite list of all "3 consecutive odd composites sandwiched between 2 primes - "triple" composites" ………. (Example: 685, 687 & 689 sandwiched between the primes 683 & 691)
(iv) <u>Element B4</u>: Infinite list of all "4 consecutive odd composites sandwiched between 2 primes - "four-ple" composites" ………. (Example: 2,769, 2,771, 2,773 & 2,775 sandwiched between the primes 2,767 & 2,777)
(v) <u>Element B5</u>: Infinite list of all "5 consecutive odd composites sandwiched between 2 primes - "five-ple" composites" ………. (Example: 19,291, 19,293, 19,295, 19,297 & 19,299 sandwiched between the primes 19,289 & 19,301)

.
.
.

(3) <u>Subgroup C</u>: Infinite consecutive odd composites separated by 4 integers and 6 integers respectively; this splitting up of the subgroup into the 2 infinite elements is shown below:

(i) <u>Element C1</u>: Infinite list of all "2 consecutive odd composites separated by 4 integers/1 prime" ………. (Example: 209 & 213 separated by the prime 211)
(ii) <u>Element C2</u>: Infinite list of all "2 consecutive odd composites separated by 6 integers/2 primes" ………. (Example: 279 & 285 separated by the twin primes 281 & 283)

(4) <u>Subgroup D</u>: Infinite single primes and twin primes separating 2 consecutive odd composites; this splitting up of the subgroup into the 2 infinite elements is shown below:

(i) <u>Element D1</u>: Infinite list of all the single primes separating 2 consecutive odd composites ………. (Example: 23 separating the 2 consecutive odd composites 21 & 25)
(ii) <u>Element D2</u>: Infinite list of all the twin primes separating 2 consecutive odd composites ………. (Example: 11 & 13 separating the 2 consecutive odd composites 9 & 15)

(5) <u>Subgroup E</u>: Infinite consecutive even composites such as 4, 6, 8, 10, 12, 14, 16, 18, 20, 22, etc. to infinity ...., all separated by only 2 integers; this subgroup may be classified as a single infinite element .... There is always 1 even number between a twin primes pair, which is separated by 2 integers, a prime and a composite which are separated by 2 integers, and, 2 composites which are separated by 2 integers. That is, the even numbers are always found in Subgroup A, Subgroup B, Subgroup C and Subgroup D above always evenly spaced out in consecutive order by 2 integers.

There is an evident symmetry in the above-mentioned infinite group, which would be broken if any of the elements within were to be finite. There are close interlinks between all the various infinite elements in all the five subgroups above, e.g., the infinity of the list of all the primes pairs each separated by 6 integers (Element A3/Subgroup A) implies the infinity of the list of all the "2 consecutive odd composites sandwiched between 2 primes - twin composites" (Element B2/Subgroup B) and vice versa, the infinity of the list of all the primes pairs each separated by 2 integers (twin primes) (Element A1/SubgroupA) implies the infinity of the list of all the "2 consecutive odd composites separated by 6 integers/2 primes" (Element C2/Subgroup C) and vice versa, the infinity of the list of all the infinite elements (A1, A2, A3, etc. to infinity) in Subgroup A above, which represents the infinity of the list of the primes which Euclid had in fact proven, implies the infinity of the list of all the "2 consecutive odd composites separated by 4 integers/1 prime" (Element C1/Subgroup C) and vice versa, the infinity of the list of all the "2 consecutive odd composites separated by 6 integers/2 primes" (Element C2/Subgroup C) implies the infinity of the list of all the twin primes separating 2 consecutive odd composites (Element D2/Subgroup D) and vice versa, the infinity of all the lists of all the infinite elements (A1, A2, A3 ......, B1, B2, B3 ......, C1 & C2, D1 & D2) in Subgroup A, Subgroup B, Subgroup C and Subgroup D above implies the infinity of the list of the consecutive even composites, i.e., 4, 6, 8, 10, 12, 14, 16, 18, 20, 22, etc. to infinity (Subgroup E), which we know to be true in any case, and vice versa, etc. to infinity.

Subgroup A and Subgroup B above are practically "mirror" images of one another - they represent the viewing of the primes and the composites from 2 variant angles - the infinitude, or, finiteness of either implies the infinitude, or, finiteness of the other; the same applies to both Subgroup C and Subgroup D above. It is similar to the following way of viewing a glass which is partially filled: this glass could be described as "half full" or "half empty" if it is half filled, "three-quarter full" or "one-quarter empty" if it is three-quarter filled, or, "one-quarter full" or "three-quarter empty" if it is one-quarter filled, etc.

It is evident that the infinitude, or, finiteness of any one of the above-mentioned elements would imply the infinitude, or, finiteness of the other element that is interlinked with it and vice versa. All these infinite elements are evidently entangled together and complementary, being all the infinite building-blocks of the infinite integers or whole numbers. The infinity of the list of the integers or whole numbers, the primes included, in fact implies that all these various elements within it should be infinite, and, vice versa, since all these various elements are closely interlinked and could not do without each other. Therefore, the breaking of the evident intrinsic symmetry of this whole infinite group, i.e., the infinite list of the integers or whole numbers, due to the finiteness of any of the elements within it, could not be possible.

We pose a very important question: Besides questioning whether the infinite list of all the primes pairs separated by 2 integers (twin primes) is really infinite, should we not also be questioning whether the following are really infinite?:

(a) Infinite lists of all the primes pairs separated respectively by 4 integers, 6 integers, 8 integers, 10 integers and sequentially larger integers to infinity (as in Subgroup A above).
(b) Infinite lists of all the respective consecutive odd composites of "infinite sizes" sandwiched between 2 primes (as in Subgroup B above).
(c) The 2 infinite lists with respectively "2 consecutive odd composites separated by 4 integers/1 prime" and "2 consecutive odd composites separated by 6 integers/2

primes" (as in Subgroup C above).
(d) The 2 infinite lists of respective single primes and twin primes separating 2 consecutive odd composites (as in Subgroup D above).
(e) Infinite list of the consecutive even composites all separated by only 2 integers (as in Subgroup E above).

Could there possibly be any symmetry-breaking in the above-mentioned infinite group whence one or more of the elements within it would be finite? In particular, could there be a possibility for the symmetry of this infinite group to be broken due to the finiteness of Element A1 (i.e., the finiteness of the twin primes) within it? Since the above-mentioned group, i.e., the list of the integers or whole numbers, is infinite, it is indeed not possible for all of these elements to be finite. And, there is no evident reason to account for why any of these elements, especially Element A1, i.e., the list of primes separated by 2 integers, or, twin primes, should be finite. In fact, all these infinite elements are like the slabs of various sizes in a building. They are all necessary for the construction of the infinite building known as the "infinite list of the integers or whole numbers" and should thus all be infinite, wherein the symmetry of the infinite group, i.e., the infinite list of the integers or whole numbers, would be preserved.

Therefore, by Lemma 1, all the elements in Subgroup A, Subgroup B, Subgroup C, Subgroup D and Subgroup E above should be infinite.

Lemma 2. The Fundamental Theorem of Arithmetic or Unique Factorisation Theorem states that there is only one possible combination of primes which will multiply together to produce any particular composite number, e.g., the only combination of primes which will produce the composite number 2,079 is: 3 x 3 x 3 x 7 x 11. In the same manner, the following composite numbers are also uniquely factorised:

(1) 63 = 3 x 3 x 7 (only)
(2) 153 = 3 x 3 x 17 (only)
(3) 1,021,020 = 2 x 2 x 3 x 5 x 7 x 11 x 13 x 17 (only)

In other words, every positive whole number which is not prime (i.e., every positive whole number which is composite) can be broken up into prime factors, and, this can happen in only one way:

$$c = \prod_{p \text{ prime}} p \quad \text{(in only one way)}$$

The 10 consecutive twin primes 3 & 5 to 107 & 109, e.g., give rise to the following 10 composite numbers which can be factorised in only one way, i.e., can be factorised only by the respective twin primes:

(1) 15 = 3 x 5 (only)
(2) 35 = 5 x 7 (only)
(3) 143 = 11 x 13 (only)
(4) 323 = 17 x 19 (only)
(5) 899 = 29 x 31 (only)
(6) 1,763 = 41 x 43 (only)
(7) 3,599 = 59 x 61 (only)
(8) 5,183 = 71 x 73 (only)
(9) 10,403 = 101 x 103 (only)
(10) 11,663 = 107 x 109 (only)
.
.

As the composite numbers are infinite, this implies that there should be an infinitude of twin primes acting as prime factors for an infinitude of composite numbers in only one way as the twin primes are indispensable, i.e., necessary, as prime factors for the formation of the composite numbers which can only be formed through the product of twin primes in only one way - the twin primes can never be substituted as prime factors of these composite numbers by other primes.

Thus, the twin primes are possibly infinite.

8

PRIMES AND EVEN NUMBERS

This chapter presents insights and many important points on the prime numbers, which are the building-blocks or "atoms" of the integers, and the Goldbach conjecture formulated by Christian Goldbach (1690 - 1764), all of which would be of interest to researchers working on the prime numbers and the Goldbach conjecture itself. The Goldbach conjecture, viz., every even number after 2 is the sum of two primes, is actually related to the distribution or "behaviour" of the prime numbers. Therefore, when the distribution or "behaviour" of the prime numbers is firmly understood the conjecture could be more easily solved. The chapter has much to share about the distribution or "behaviour" of the prime numbers, providing much numerical evidence to support the conjecture, besides suggesting ways or arguments for resolving the conjecture.

Introduction

The expected mode of solving the Goldbach conjecture appears to be the utilisation of advanced calculus or analysis, e.g., by the summation, or, integration, of the reciprocals involving directly or indirectly the primes to see whether they converge or diverge, in order to get a "feel" of the pattern of the distribution of the primes. But, such a method of solving the problem has evidently not succeeded so far. Some other approach or approaches could be more appropriate. This chapter addresses the problem from several different angles, with reasoning backed by quantities that can be checked. Importantly, it explains the link between the distribution or "behaviour" of the prime numbers and the Goldbach conjecture; in fact the validity of Goldbach conjecture is a function of the distribution or "behaviour" of the prime numbers.

Primes and the Even Numbers

The following chain of reasoning explains the link between the primes and the Goldbach conjecture and could possibly validate the conjecture:-

Every even number after 2 is the sum of 2 odd numbers. Every odd number is either a prime which is odd or a composite - product of primes which are odd; notably, every prime with the exception of 2 is an odd number. Every even number after 2 is also a composite, but, a composite with at least 1 even prime factor, namely, 2, while the rest of its prime factors are odd, i.e., it is an even composite.

Therefore, every even number after 2 is the sum of 2 primes which are odd and/or the sum of 1 prime which is odd and 1 odd composite whose prime factors are odd and/or the sum of 2 odd composites whose prime factors are odd, besides being an even composite with at least 1 even prime factor, namely, 2, while the rest of its prime factors are odd.

Lemma. By Euclid's theorem, the primes are infinite; this implies that there would be an infinitude of sums of 2 primes as per the Goldbach conjecture. The even numbers, which are sums of 2 primes as per the conjecture, are also infinite. Thus, there are an infinite number of even numbers which are sums of 2 primes, both the even numbers and sums of 2 primes being infinite.

Corollary. The odd numbers, which are either prime, every prime with the exception of 2 being an odd number, or composite (have prime factors which are odd), are infinite; this implies that there would be an infinite number of sums of 2 odd numbers, each of which is equal to an even number. Hence, as there is an infinitude of even numbers which are sums of 2 primes, as per the above lemma, and as all primes with the exception of 2 are odd numbers, there are an infinite number of even numbers which are sums of 2 odd numbers that are prime, all the even numbers, sums of 2 odd numbers and primes being infinite; i.e., every even number after 2 appears to be the sum of 2 odd numbers that are prime.

We thereby see the close interlink or relationship between the primes, even numbers and odd numbers, which are all infinite, which is significant.

The following are thus evident:

a) Every sum of 2 primes which are odd numbers is equal to an even number, as is below in consecutive order:

$2 + 2 = 1 + 3 = \mathbf{4}$
$3 + 3 = 1 + 5 = \mathbf{6}$
$3 + 5 = 1 + 7 = \mathbf{8}$
$5 + 5 = 3 + 7 = \mathbf{10}$
$5 + 7 = 1 + 11 = \mathbf{12}$
$7 + 7 = 3 + 11 = 1 + 13 = \mathbf{14}$
$3 + 13 = 5 + 11 = \mathbf{16}$
$7 + 11 = 5 + 13 = 1 + 17 = \mathbf{18}$
$7 + 13 = 3 + 17 = 1 + 19 = \mathbf{20}$
$11 + 11 = 3 + 19 = 5 + 17 = 11 + 11 = \mathbf{22}$
$11 + 13 = 5 + 19 = 7 + 17 = 1 + 23 = \mathbf{24}$
$13 + 13 = 3 + 23 = 7 + 19 = \mathbf{26}$
$11 + 17 = 5 + 23 = \mathbf{28}$
$13 + 17 = 11 + 19 = 7 + 23 = 1 + 29 = \mathbf{30}$
$3 + 29 = 13 + 19 = 1 + 31 = \mathbf{32}$
$17 + 17 = 3 + 31 = 5 + 29 = 11 + 23 = 17 + 17 = \mathbf{34}$
$17 + 19 = 5 + 31 = 7 + 29 = 13 + 23 = \mathbf{36}$
$19 + 19 = 7 + 31 = 1 + 37 = \mathbf{38}$
$3 + 37 = 11 + 29 = 17 + 23 = \mathbf{40}$
$19 + 23 = 5 + 37 = 11 + 31 = 13 + 29 = 1 + 41 = \mathbf{42}$
$3 + 41 = 7 + 37 = 13 + 31 = 1 + 43 = \mathbf{44}$
$23 + 23 = 3 + 43 = 5 + 41 = 17 + 29 = \mathbf{46}$
$5 + 43 = 7 + 41 = 11 + 37 = 17 + 31 = 19 + 29 = 1 + 47 = \mathbf{48}$
$3 + 47 = 7 + 43 = 13 + 37 = 19 + 31 = \mathbf{50}$
$23 + 29 = 5 + 47 = 11 + 41 = \mathbf{52}$
$7 + 47 = 11 + 43 = 13 + 41 = 17 + 37 = 23 + 31 = 1 + 53 = \mathbf{54}$
$3 + 53 = 13 + 43 = 19 + 37 = \mathbf{56}$
$29 + 29 = 5 + 53 = 11 + 47 = 17 + 41 = 29 + 29 = \mathbf{58}$
$29 + 31 = 7 + 53 = 13 + 47 = 17 + 43 = 19 + 41 = 23 + 37 = 1 + 59 = \mathbf{60}$
$31 + 31 = 3 + 59 = 19 + 43 = 1 + 61 = \mathbf{62}$
$3 + 61 = 5 + 59 = 11 + 53 = 17 + 47 = 23 + 41 = \mathbf{64}$
$5 + 61 = 7 + 59 = 13 + 53 = 19 + 47 = 23 + 43 = 29 + 37 = \mathbf{66}$
$7 + 61 = 31 + 37 = 1 + 67 = \mathbf{68}$
$3 + 67 = 11 + 59 = 17 + 53 = 23 + 47 = 29 + 41 = \mathbf{70}$
$5 + 67 = 11 + 61 = 13 + 59 = 19 + 53 = 29 + 43 = 31 + 41 = 1 + 71 = \mathbf{72}$
$37 + 37 = 3 + 71 = 7 + 67 = 13 + 61 = 31 + 43 = 37 + 37 = 1 + 73 = \mathbf{74}$

3 + 73 = 5 + 71 = 17 + 59 = 23 + 53 = 29 + 47 = **76**
37 + 41 = 5 + 73 = 7 + 71 = 11 + 67 = 31 + 47 = 37 + 41 = **78**
7 + 73 = 13 + 67 = 19 + 61 = 37 + 43 = 1 + 79 = **80**
41 + 41 = 3 + 79 = 11 + 71 = 23 + 59 = 29 + 53 = **82**
41 + 43 = 5 + 79 = 11 + 73 = 13 + 71 = 17 + 67 = 23 + 61 = 31 + 53 = 37 + 47 = 1+ 83 = **84**
43 + 43 = 3 + 83 = 7 + 79 = 13 + 73 = 19 + 67 = 43 + 43 = **86**
5 + 83 = 17 + 71 = 29 + 59 = 41 + 47 = **88**
7 + 83 = 11 + 79 = 17 + 73 = 19 + 71 = 23 + 67 = 29 + 61 = 31 + 59 = 37 + 53 = 43 + 47 = 1 + 89 = **90**
3 + 89 = 13 + 79 = 19 + 73 = 31 + 61 = 1 + 91 = **92**
47 + 47 = 5 + 89 = 11 + 83 = 23 + 71 = 41 + 53 = 47 + 47 = **94**
5 + 91 = 7 + 89 = 13 + 83 = 17 + 79 = 23 + 73 = 29 + 67 = 37 + 59 = 43 + 53 = **96**
7 + 91 = 19 + 79 = 31 + 67 = 37 + 61 = 1 + 97 = **98**
47 + 53 = 3 + 97 = 11 + 89 = 17 + 83 = 29 + 71 = 41 + 59 = 47 + 53 = **100**
5 + 97 = 11 + 91 = 13 + 89 = 19 + 83 = 23 + 79 = 29 + 73 = 31 + 71 = 41 + 61 = 43 + 59 = 1 + 101 = **102**

.
.
.

b) Every sum of 1 prime which is an odd number & 1 odd composite which is the product of primes which are odd, is equal to the sum of 2 primes which are odd numbers, which are all each equal to an even number, as is below in consecutive order:

3 + 9 = 5 + 7 = 1 + 11 = **12**
5 + 9 = 3 + 11 = 7 + 7 = 1 + 13 = **14**
7 + 9 = 3 + 13 = 5 + 11 = **16**
3 + 15 = 7 + 11 = 5 + 13 = 1 + 17 = **18**
11 + 9 = 3 + 17 = 7 + 13 = 1 + 19 = **20**
13 + 9 = 3 + 19 = 5 + 17 = 11 + 11 = **22**
3 + 21 = 11 + 13 = 5 + 19 = 7 + 17 = 1 + 23 = **24**
17 + 9 = 3 + 23 = 7 + 19 = 13 + 13 = **26**
19 + 9 = 5 + 23 = 11 + 17 = **28**
5 + 25 = 13 + 17 = 11 + 19 = 7 + 23 = 1 + 29 = **30**
23 + 9 = 3 + 29 = 13 + 19 = 1 + 31 = **32**
7 + 27 = 17 + 17 = 3 + 31 = 5 + 29 = 11 + 23 = 17 + 17 = **34**
3 + 33 = 17 + 19 = 5 + 31 = 7 + 29 = 13 + 23 = **36**
29 + 9 = 7 + 31 = 19 + 19 = 1 + 37 = **38**
31 + 9 = 3 + 37 = 11 + 29 = 17 + 23 = **40**
3 + 39 = 19 + 23 = 5 + 37 = 11 + 31 = 13 + 29 = 1 + 41 = **42**
5 + 39 = 3 + 41 = 7 + 37 = 13 + 31 = 1 + 43 = **44**
37 + 9 = 3 + 43 = 5 + 41 = 17 + 29 = 23 + 23 = **46**
3 + 45 = 5 + 43 = 7 + 41 = 11 + 37 = 17 + 31 = 19 + 29 = 1 + 47 = **48**
41 + 9 = 3 + 47 = 7 + 43 = 13 + 37 = 19 + 31 = **50**
43 + 9 = 5 + 47 = 11 + 41 =23 + 29 = **52**
5 + 49 = 7 + 47 = 11 + 43 = 13 + 41 = 17 + 37 = 23 + 31 = 1 + 53 = **54**
47 + 9 = 3 + 53 = 13 + 43 = 19 + 37 = **56**
3 + 55 = 29 + 29 = 5 + 53 = 11 + 47 = 17 + 41 = 29 + 29 = **58**

5 + 55 = 29 + 31 = 7 + 53 = 13 + 47 = 17 + 43 = 19 + 41 = 23 + 37 = 1 + 59 = **60**
53 + 9 = 3 + 59 = 19 + 43 = 31 + 31 = 1 + 61 = **62**
7 + 57 = 3 + 61 = 5 + 59 = 11 + 53 = 17 + 47 = 23 + 41 = **64**
11 + 55 = 5 + 61 = 7 + 59 = 13 + 53 = 19 + 47 = 23 + 43 = 29 + 37 = **66**
59 + 9 = 7 + 61 = 31 + 37 = 1 + 67 = **68**
61 + 9 = 3 + 67 = 11 + 59 = 17 + 53 = 23 + 47 = 29 + 41 = **70**
3 + 69 = 5 + 67 = 11 + 61 = 13 + 59 = 19 + 53 = 29 + 43 = 31 + 41 = 1 + 71 = **72**
5 + 69 = 37 + 37 = 3 + 71 = 7 + 67 = 13 + 61 = 31 + 43 = 37 + 37 = 1 + 73 = **74**
67 + 9 = 3 + 73 = 5 + 71 = 17 + 59 = 23 + 53 = 29 + 47 = **76**
3 + 75 = 37 + 41 = 5 + 73 = 7 + 71 = 11 + 67 = 31 + 47 = 37 + 41 = **78**
71 + 9 = 7 + 73 = 13 + 67 = 19 + 61 = 37 + 43 = 1 + 79 = **80**
73 + 9 = 3 + 79 = 11 + 71 = 23 + 59 = 29 + 53 = 41 + 41 = **82**
3 + 81 = 41 + 43 = 5 + 79 = 11 + 73 = 13 + 71 = 17 + 67 = 23 + 61 = 31 + 53 = 37 + 47 = 1+ 83 = **84**
5 + 81 = 43 + 43 = 3 + 83 = 7 + 79 = 13 + 73 = 19 + 67 = 43 + 43 = **86**
79 + 9 = 5 + 83 = 17 + 71 = 29 + 59 = 41 + 47 = **88**
3 + 87 = 7 + 83 = 11 + 79 = 17 + 73 = 19 + 71 = 23 + 67 = 29 + 61 = 31 + 59 = 37 + 53 = 43 + 47 = 1 + 89 = **90**
83 + 9 = 3 + 89 = 13 + 79 = 19 + 73 = 31 + 61 = 1 + 91 = **92**
7 + 87 = 47 + 47 = 5 + 89 = 11 + 83 = 23 + 71 = 41 + 53 = 47 + 47 = **94**
3 + 93 = 5 + 91 = 7 + 89 = 13 + 83 = 17 + 79 = 23 + 73 = 29 + 67 = 37 + 59 = 43 + 53 = **96**
89 + 9 = 7 + 91 = 19 + 79 = 31 + 67 = 37 + 61 = 1 + 97 = **98**
91 + 9 = 3 + 97 = 11 + 89 = 17 + 83 = 29 + 71 = 41 + 59 = 47 + 53 = **100**
3 + 99 = 5 + 97 = 11 + 91 = 13 + 89 = 19 + 83 = 23 + 79 = 29 + 73 = 31 + 71 = 41 + 61 = 43 + 59 = 1 + 101 = **102**

.
.
.

c) Every sum of 2 odd composites which are products of primes which are odd, is equal to the sum of 2 primes which are odd numbers, which are all each equal to an even number, as is below in consecutive order:

9 + 9 = 5 + 13 = 7 + 11 = 1 + 17 = **18**
9 + 15 = 5 + 19 = 7 + 17 = 11 + 13 = 1 + 23 = **24**
15 + 15 = 7 + 23 = 11 + 19 = 13 + 17 = 1 + 29 = **30**
9 + 25 = 7 + 27 = 17 + 17 = 3 + 31 = 5 + 29 = 11 + 23 = 17 + 17 = **34**
15 + 21 = 5 + 31 = 7 + 29 = 13 + 23 = 17 + 19 = **36**
15 + 25 = 3 + 37 = 11 + 29 = 17 + 23 = **40**
21 + 21 = 5 + 37 = 11 + 31 = 13 + 29 = 19 + 23 = 1 + 41 = **42**
9 + 35 = 3 + 41 = 7 + 37 = 13 + 31 = 1 + 43 = **44**
21 + 25 = 3 + 43 = 5 + 41 = 17 + 29 = 23 + 23 = **46**
9 + 39 = 5 + 43 = 7 + 41 = 11 + 37 = 17 + 31 = 19 + 29 = 1 + 47 = **48**
25 + 25 = 3 + 47 = 7 + 43 = 13 + 37 = 19 + 31 = **50**
25 + 27 = 5 + 47 = 11 + 41 =23 + 29 = **52**
27 + 27 = 7 + 47 = 11 + 43 = 13 + 41 = 17 + 37 = 23 + 31 = 1 + 53 = **54**
21 + 35 = 3 + 53 = 13 + 43 = 19 + 37 = **56**
9 + 49 = 29 + 29 = 5 + 53 = 11 + 47 = 17 + 41 = 29 + 29 = **58**
27 + 33 = 7 + 53 = 13 + 47 = 17 + 43 = 19 + 41 = 23 + 37 = 29 + 31 = 1 + 59 = **60**
27 + 35 = 31 + 31 = 3 + 59 = 19 + 43 = 1 + 61 = **62**

9 + 55 = 3 + 61 = 5 + 59 = 11 + 53 = 17 + 47 = 23 + 41 = **64**
33 + 33 = 5 + 61 = 7 + 59 = 13 + 53 = 19 + 47 = 23 + 43 = 29 + 37 = **66**
33 + 35 = 7 + 61 = 31 + 37 = 1 + 67 = **68**
35 + 35 = 3 + 67 = 11 + 59 = 17 + 53 = 23 + 47 = 29 + 41 = **70**
9 + 63 = 5 + 67 = 11 + 61 = 13 + 59 = 19 + 53 = 29 + 43 = 31 + 41 = 1 + 71 = **72**
35 + 39 = 3 + 71 = 7 + 67 = 13 + 61 = 31 + 43 = 37 + 37 = 1 + 73 = **74**
21 + 55 = 3 + 73 = 5 + 71 = 17 + 59 = 23 + 53 = 29 + 47 = **76**
39 + 39 = 5 + 73 = 7 + 71 = 11 + 67 = 31 + 47 = 37 + 41 = **78**
15 + 65 = 7 + 73 = 13 + 67 = 19 + 61 = 37 + 43 = 1 + 79 = **80**
25 + 57 = 41 + 41 = 3 + 79 = 11 + 71 = 23 + 59 = 29 + 53 = **82**
39 + 45 = 5 + 79 = 11 + 73 = 13 + 71 = 17 + 67 = 23 + 61 = 31 + 53 = 37 + 47 = 41 + 43 = 1 + 83 = **84**
9 + 77 = 43 + 43 = 3 + 83 = 7 + 79 = 13 + 73 = 19 + 67 = 43 + 43 = **86**
25 + 63 = 5 + 83 = 17 + 71 = 29 + 59 = 41 + 47 = **88**
45 + 45 = 7 + 83 = 11 + 79 = 17 + 73 = 19 + 71 = 23 + 67 = 29 + 61 = 31 + 59 = 37 + 53 = 43 + 47 = 1 + 89 = **90**
15 + 77 = 3 + 89 = 13 + 79 = 19 + 73 = 31 + 61 = 1 + 91 = **92**
45 + 49 = 5 + 89 = 11 + 83 = 23 + 71 = 41 + 53 = 47 + 47 = **94**
9 + 87 = 5 + 91 = 7 + 89 = 13 + 83 = 17 + 79 = 23 + 73 = 29 + 67 = 37 + 59 = 43 + 53 = **96**
49 + 49 = 7 + 91 = 19 + 79 = 31 + 67 = 37 + 61 = 1 + 97 = **98**
49 + 51 = 3 + 97 = 11 + 89 = 17 + 83 = 29 + 71 = 41 + 59 = 47 + 53 = **100**
51 + 51 = 5 + 97 = 11 + 91 = 13 + 89 = 19 + 83 = 23 + 79 = 29 + 73 = 31 + 71 = 41 + 61 = 43 + 59 = 1 + 101 = **102**

.
.
.

d) From (a), (b) & (c) above, we have the even numbers from 4 to 102 … composed as follows:

1) **4** = 2 + 2 = 1 + 3 (sum of 2 primes only)
2) **6** = 3 + 3 = 1 + 5 (sum of 2 primes only)
3) **8** = 3 + 5 = 1 + 7 (sum of 2 primes only)
4) **10** = 5 + 5 = 3 + 7 (sum of 2 primes only)
5) **12** = 5 + 7 = 1 + 11 = **3 + 9** (sum of 1 prime & 1 odd composite)
6) **14** = 3 + 11 = 7 + 7 = 1 + 13 = **5 + 9** (sum of 1 prime & 1 odd composite)
7) **16** = 3 + 13 = 5 + 11 = **7 + 9** (sum of 1 prime & 1 odd composite)
8) **18** = 5 + 13 = 7 + 11 = 1 + 17 = **3 + 15** (sum of 1 prime & 1 odd composite) = **9 + 9** (sum of 2 odd composites)
9) **20** = 3 + 17 = 7 + 13 = 1 + 19 = **11 + 9** (sum of 1 prime & 1 odd composite)
10) **22** = 3 + 19 = 5 + 17 = 11 + 11 = **13 + 9** (sum of 1 prime & 1 odd composite)
11) **24** = 5 + 19 = 7 + 17 = 11 + 13 = 1 + 23 = **3 + 21** (sum of 1 prime & 1 odd composite) = **9 + 15** (sum of 2 odd composites)
12) **26** = 3 + 23 = 7 + 19 = 13 + 13 = **17 + 9** (sum of 1 prime & 1 odd composite)
13) **28** = 5 + 23 = 11 + 17 = **19 + 9** (sum of 1 prime & 1 odd composite)
14) **30** = 7 + 23 = 11 + 19 = 13 + 17 = 1 + 29 = **5 + 25** (sum of 1 prime & 1 odd composite) = **15 + 15** (sum of 2 odd composites)
15) **32** = 3 + 29 = 13 + 19 = 1 + 31 = **23 + 9** (sum of 1 prime & 1 odd composite)
16) **34** = 17 + 17 = 3 + 31 = 5 + 29 = 11 + 23 = 17 + 17 = **7 + 27** (sum of 1 prime & 1 odd composite) = **9 + 25** (sum of 2 odd composites)

17) **36** = 5 + 31 = 7 + 29 = 13 + 23 = 17 + 19 = **3 + 33** (sum of 1 prime & 1 odd composite) = **15 + 21** (sum of 2 odd composites)
18) **38** = 7 + 31 = 19 + 19 = 1 + 37 = **29 + 9** (sum of 1 prime & 1 odd composite)
19) **40** = 3 + 37 = 11 + 29 = 17 + 23 = **31 + 9** (sum of 1 prime & 1 odd composite) = **15 + 25** (sum of 2 odd composites)
20) **42** = 5 + 37 = 11 + 31 = 13 + 29 = 19 + 23 = 1 + 41 = **3 + 39** (sum of 1 prime & 1 odd composite) = **21 + 21** (sum of 2 odd composites)
21) **44** = 3 + 41 = 7 + 37 = 13 + 31 = 1 + 43 = **5 + 39** (sum of 1 prime & 1 odd composite) = **9 + 35** (sum of 2 odd composites)
22) **46** = 3 + 43 = 5 + 41 = 17 + 29 = 23 + 23 = **37 + 9** (sum of 1 prime & 1 odd composite) = **21 + 25** (sum of 2 odd composites)
23) **48** = 5 + 43 = 7 + 41 = 11 + 37 = 17 + 31 = 19 + 29 = 1 + 47 = **3 + 45** (sum of 1 prime & 1 odd composite) = **9 + 39** (sum of 2 odd composites)
24) **50** = 3 + 47 = 7 + 43 = 13 + 37 = 19 + 31 = **41 + 9** (sum of 1 prime & 1 odd composite) = **25 + 25** (sum of 2 odd composites)
25) **52** = 5 + 47 = 11 + 41 = 23 + 29 = = **43 + 9** (sum of 1 prime & 1 odd composite) = **25 + 27** (sum of 2 odd composites)
26) **54** = 7 + 47 = 11 + 43 = 13 + 41 = 17 + 37 = 23 + 31 = 1 + 53 = **5 + 49** (sum of 1 prime & 1 odd composite) = **27 + 27** (sum of 2 odd composites)
27) **56** = 3 + 53 = 13 + 43 = 19 + 37 = **47 + 9** (sum of 1 prime & 1 odd composite) = **21 + 35** (sum of 2 odd composites)
28) **58** = 29 + 29 = 5 + 53 = 11 + 47 = 17 + 41 = 29 + 29 = **3 + 55** (sum of 1 prime & 1 odd composite) = **9 + 49** (sum of 2 odd composites)
29) **60** = 7 + 53 = 13 + 47 = 17 + 43 = 19 + 41 = 23 + 37 = 29 + 31 = 1 + 59 = **5 + 55** (sum of 1 prime & 1 odd composite) = **27 + 33** (sum of 2 odd composites)
30) **62** = 3 + 59 = 19 + 43 = 31 + 31 = 1 + 61 = **53 + 9** (sum of 1 prime & 1 odd composite) = **27 + 35** (sum of 2 odd composites)
31) **64** = 3 + 61 = 5 + 59 = 11 + 53 = 17 + 47 = 23 + 41 = **7 + 57** (sum of 1 prime & 1 odd composite) = **9 + 55** (sum of 2 odd composites)
32) **66** = 5 + 61 = 7 + 59 = 13 + 53 = 19 + 47 = 23 + 43 = 29 + 37 = **11 + 55** (sum of 1 prime & 1 odd composite) = **33 + 33** (sum of 2 odd composites)
33) **68** = 7 + 61 = 31 + 37 = 1 + 67 = **59 + 9** (sum of 1 prime & 1 odd composite) = **33 + 35** (sum of 2 odd composites)
34) **70** = 3 + 67 = 11 + 59 = 17 + 53 = 23 + 47 = 29 + 41 = **61 + 9** (sum of 1 prime & 1 odd composite) = **35 + 35** (sum of 2 odd composites)
35) **72** = 5 + 67 = 11 + 61 = 13 + 59 = 19 + 53 = 29 + 43 = 31 + 41 = 1 + 71 = **3 + 69** (sum of 1 prime & 1 odd composite) = **9 + 63** (sum of 2 odd composites)
36) **74** = 3 + 71 = 7 + 67 = 13 + 61 = 31 + 43 = 37 + 37 = 1 + 73 = **5 + 69** (sum of 1 prime & 1 odd composite) = **35 + 39** (sum of 2 odd composites)
37) **76** = 3 + 73 = 5 + 71 = 17 + 59 = 23 + 53 = 29 + 47 = **67 + 9** (sum of 1 prime & 1 odd composite) = **21 + 55** (sum of 2 odd composites)
38) **78** = 5 + 73 = 7 + 71 = 11 + 67 = 31 + 47 = 37 + 41 = **3 + 75** (sum of 1 prime & 1 odd composite) = **39 + 39** (sum of 2 odd composites)
39) **80** = 7 + 73 = 13 + 67 = 19 + 61 = 37 + 43 = 1 + 79 = **71 + 9** (sum of 1 prime & 1 odd composite) = **15 + 65** (sum of 2 odd composites)
40) **82** = 3 + 79 = 11 + 71 = 23 + 59 = 29 + 53 = 41 + 41 = **73 + 9** (sum of 1 prime & 1 odd composite) = **25 + 57** (sum of 2 odd composites)
41) **84** = 5 + 79 = 11 + 73 = 13 + 71 = 17 + 67 = 23 + 61 = 31 + 53 = 37 + 47 = 41 + 43 = 1 + 83 = **3 + 81** (sum of 1 prime & 1 odd composite) = **39 + 45** (sum of 2 odd composites)
42) **86** = 43 + 43 = 3 + 83 = 7 + 79 = 13 + 73 = 19 + 67 = 43 + 43 = **5 + 81** (sum of

1 prime & 1 odd composite) = **9 + 77** (sum of 2 odd composites)
43) **88** = 5 + 83 = 17 + 71 = 29 + 59 = 41 + 47 = **79 + 9** (sum of 1 prime & 1 odd composite) = **25 + 63** (sum of 2 odd composites)
44) **90** = 7 + 83 = 11 + 79 = 17 + 73 = 19 + 71 = 23 + 67 = 29 + 61 = 31 + 59 = 37 + 53 = 43 + 47 = 1 + 89 = **3 + 87** (sum of 1 prime & 1 odd composite) = **45 + 45** (sum of 2 odd composites)
45) **92** = 3 + 89 = 13 + 79 = 19 + 73 = 31 + 61 = 1 + 91 = **83 + 9** (sum of 1 prime & 1 odd composite) = **15 + 77** (sum of 2 odd composites)
46) **94** = 5 + 89 = 11 + 83 = 23 + 71 = 41 + 53 = 47 + 47 = **7 + 87** (sum of 1 prime & 1 odd composite) = **45 + 49** (sum of 2 odd composites)
47) **96** = 5 + 91 = 7 + 89 = 13 + 83 = 17 + 79 = 23 + 73 = 29 + 67 = 37 + 59 = 43 + 53 = **3 + 93** (sum of 1 prime & 1 odd composite) = **9 + 87** (sum of 2 odd composites)
48) **98** = 7 + 91 = 19 + 79 = 31 + 67 = 37 + 61 = 1 + 97 = **89 + 9** (sum of 1 prime & 1 odd composite) = **49 + 49** (sum of 2 odd composites)
49) **100** = 3 + 97 = 11 + 89 = 17 + 83 = 29 + 71 = 41 + 59 = 47 + 53 = **91 + 9** (sum of 1 prime & 1 odd composite) = **49 + 51** (sum of 2 odd composites)
50) **102** = 5 + 97 = 11 + 91 = 13 + 89 = 19 + 83 = 23 + 79 = 29 + 73 = 31 + 71 = 41 + 61 = 43 + 59 = 1 + 101 = **3 + 99** (sum of 1 prime & 1 odd composite) = **51 + 51** (sum of 2 odd composites)

.
.
.

(The above is only a partial or incomplete listing of sums of 1 prime & 1 odd composite, and, sums of 2 odd composites, each of which is equal to the sum of 2 primes as well as an even number. For example, in the list of compositions for the even numbers 4 to 102 … above, in Item (48), we could also have other "combinations" such as: **98** = 7 + 91 = 19 + 79 = 31 + 67 = 37 + 61 = 1 + 97 = **25 + 73** (sum of 1 prime & 1 odd composite) = **21 + 77** (sum of 2 odd composites), etc., in Item (49), we could also have other "combinations" such as: **100** = 3 + 97 = 11 + 89 = 17 + 83 = 29 + 71 = 41 + 59 = 47 + 53 = **31 + 69** (sum of 1 prime & 1 odd composite) = **45 + 55** (sum of 2 odd composites), etc., and, in Item (50), we could also have other "combinations" such as: **102** = 5 + 97 = 11 + 91 = 13 + 89 = 19 + 83 = 23 + 79 = 29 + 73 = 31 + 71 = 41 + 61 = 43 + 59 = 1 + 101 = **17 + 85** (sum of 1 prime & 1 odd composite) = **21 + 81** (sum of 2 odd composites), etc. That is, there are more "combinations" than those shown in the above listing.)

In (d) above, in the list of compositions for the 50 consecutive even numbers 4 to 102 …, the even numbers 4, 6, 8 and 10 are only formed through the summing of 2 primes and not at all through the summing of 1 prime and 1 odd composite, or, the summing of 2 odd composites, which are impossibilities here. These sums of 2 primes are present (always present) throughout the whole list of compositions, from 4 right through to 102, while this is not the case for the sums of 1 prime and 1 odd composite, and, the sums of 2 odd composites.

We reason here by the process of elimination, through analysing the information in (d) above which pertains to the compositions of the 50 consecutive even numbers 4 to 102 … taken from the infinite list of even numbers. We stated at the beginning the following about the even numbers after 2:-

Firstly, every even number after 2 is:
A) The sum of 2 odd numbers.
(Every odd number is either a prime which is odd or a composite - product of primes which are odd.
Notably, every prime with the exception of 2 is an odd number.)

Secondly, every even number after 2 is also (the below-mentioned is the logical consequence of (A) above):
1) The sum of 2 primes which are odd.
2) And/or the sum of 1 prime which is odd and 1 odd composite whose prime factors are odd.
3) And/or the sum of 2 odd composites whose prime factors are odd.

Evidently, at least 1 of (1), (2) & (3) above has to be the "atom" or building-block of the even numbers. In (d) above, we observe the following:-

i) All the 50 consecutive even numbers 4 to 102 ... in (d) above taken from the infinite list of even numbers are sums of 2 primes.
ii) It is impossible for each of the even numbers 4, 6, 8 & 10 in (d) above to be the sum of 1 prime which is odd and 1 odd composite whose prime factors are odd.
iii) It is impossible for each of the even numbers 4, 6, 8, 10, 12, 14, 16, 20, 22, 26, 28, 32 & 38 in (d) above to be the sum of 2 odd composites whose prime factors are odd.

It is evident from (i), (ii) & (iii) above that neither (2) nor (3) can be the "atom" or building-block of the even numbers since they are "incomplete". As (1) - the sum of 2 primes which are odd - is "complete", i.e., always present in the 50 consecutive even numbers 4 to 102 ... in (d) above, unlike (2) & (3), it evidently is the "atom" or building-block of the even numbers. That is, every even number after 2 is evidently the sum of 2 primes which are odd. In fact, a distributed computer search completed in 2008 at the University of Aveiro, Portugal, had verified this for all even numbers up to 12×10^{17}, which is not a small list (it is in fact a long, impressive list, obtainable only with the help of modern computer technology). Definitely, due respectively to (ii) & (iii) above, we cannot say that every even number after 2 is the sum of 1 prime which is odd and 1 odd composite whose prime factors are odd, or, every even number after 2 is the sum of 2 odd composites whose prime factors are odd.

By the above lemma and corollary, the infinitudes of the primes, even numbers and odd numbers indeed imply that there are an infinite number of sums of 2 primes which are odd numbers, which are each equal to an even number. As the sums of 2 primes which are odd numbers are evidently the "atoms" or building-blocks of the even numbers, it also implies that they are infinite, since the even numbers are infinite.

Hypothetically, if on the other hand just 1 of the 3 items stated above, primes, even numbers and odd numbers, were finite, the above-said sums of 2 primes which are odd numbers, each of which is equal to an even number, would be finite. The primes, even numbers and odd numbers are evidently intricately linked, with the primes playing the part of building-blocks of both the even and odd numbers through various "combinations" as is described below. However, as the primes, even numbers and odd numbers are intricately linked, the finiteness (or, infinity) of any 1 of them implies the finiteness (or, infinity) of the other 2, and vice versa. These 3 items are evidently "close comrades-in-arm" working together to give special meaning to the integers. As these 3 are all infinite, it indeed implies that there is an infinitude of even numbers which are infinitely the sums of 2 primes that are odd and infinite.

The prime numbers are evidently the atoms or building-blocks of the integers. The integers are either primes (not divisible by other integers except 1) or composites (divisible by other integers, e.g., the prime numbers), and, even (the sums of 2 primes as conjectured by Goldbach) or odd (primes, or, composites whereby they are divisible by prime factors). Therefore, to determine whether the conjecture that every even number (except the number 2) is the sum of 2 primes is true, it would be appropriate to analyse the evident atoms or building-blocks of the even numbers, viz., the prime numbers. For the solution to this conjecture we note that the primes (vide Euclid's theorem) and the even numbers are infinite, which implies that this conjecture should be true.

We here analyse the "behaviour" of the first 2,400 consecutive prime numbers (divided into 12 batches of consecutive primes, each subsequent batch with an increment of 200 primes), leaving out 2 (because it is an even prime) and commencing with 3, which is the $2^{nd.}$ consecutive prime, the latter to be the first prime in our list of 2,400 consecutive primes (3 to 21,391), as follows:-

(1) <u>200 Consecutive Primes From 3 To 1,229</u>
    (a) Even numbers (obtained by summing of 2 primes) = 6 to 2,458
    (b) No. of even numbers = 1,227
    (c) No. of primes = 200
    (d) Average no. of even numbers "generated" by each of these 200 consecutive primes = 1,227 ÷ 200 = **6.14**
    (e) No. of summings of 2 primes/permutations (3 + 3, 3 + 5, 3 + 7, 3 + 11, … etc.) for these 200 primes = 200 x 200 = 40,000
    (f) Average no. of summings of 2 primes/permutations for each of the 1,227 even numbers = 40,000 ÷ 1,227 = **32.60**

(2) <u>400 Consecutive Primes From 3 To 2,749</u>
    (a) Even numbers (obtained by summing of 2 primes) = 6 to 5,498
    (b) No. of even numbers = 2,747
    (c) No. of primes = 400
    (d) Average no. of even numbers "generated" by each of these 400 consecutive primes = 2,747 ÷ 400 = **6.87**
    (e) No. of summings of 2 primes/permutations (3 + 3, 3 + 5, 3 + 7, 3 + 11, … etc.) for these 400 primes = 400 x 400 = 160,000
    (f) Average no. of summings of 2 primes/permutations for each of the 2,747 even numbers = 160,000 ÷ 2,747 = **58.25**

(3) <u>600 Consecutive Primes From 3 To 4,421</u>
    (a) Even numbers (obtained by summing of 2 primes) = 6 to 8,842
    (b) No. of even numbers = 4,419
    (c) No. of primes = 600
    (d) Average no. of even numbers "generated" by each of these 600 consecutive primes = 4,419 ÷ 600 = **7.37**
    (e) No. of summings of 2 primes/permutations (3 + 3, 3 + 5, 3 + 7, 3 + 11, … etc.) for these 600 primes = 600 x 600 = 360,000
    (f) Average no. of summings of 2 primes/permutations for each of the 4,419 even numbers = 360,000 ÷ 4,419 = **81.47**

(4) <u>800 Consecutive Primes From 3 To 6,143</u>
    (a) Even numbers (obtained by summing of 2 primes) = 6 to 12,286
    (b) No. of even numbers = 6,141
    (c) No. of primes = 800
    (d) Average no. of even numbers "generated" by each of these 800 consecutive primes = 6,141 ÷ 800 = **7.68**
    (e) No. of summings of 2 primes/permutations (3 + 3, 3 + 5, 3 + 7, 3 + 11, … etc.) for these 800 primes = 800 x 800 = 640,000
    (f) Average no. of summings of 2 primes/permutations for each of the 6,141 even numbers = 640,000 ÷ 6,141 = **104.22**

(5) <u>1,000 Consecutive Primes From 3 To 7,927</u>
    (a) Even numbers (obtained by summing of 2 primes) = 6 to 15,854

(b) No. of even numbers = 7,925
(c) No. of primes = 1,000
(d) Average no. of even numbers "generated" by each of these 1,000 consecutive primes = 7,925 ÷ 1,000 = **7.93**
(e) No. of summings of 2 primes/permutations (3 + 3, 3 + 5, 3 + 7, 3 + 11, ... etc.) for these 1,000 primes = 1,000 x 1,000 = 1,000,000
(f) Average no. of summings of 2 primes/permutations for each of the 7,925 even numbers = 1,000,000 ÷ 7,925 = **126.18**

(6) <u>1,200 Consecutive Primes From 3 To 9,739</u>
(a) Even numbers (obtained by summing of 2 primes) = 6 to 19,478
(b) No. of even numbers = 9,737
(c) No. of primes = 1,200
(d) Average no. of even numbers "generated" by each of these 1,200 consecutive primes = 9,737 ÷ 1,200 = **8.11**
(e) No. of summings of 2 primes/permutations (3 + 3, 3 + 5, 3 + 7, 3 + 11, ... etc.) for these 1,200 primes = 1,200 x 1,200 = 1,440,000
(f) Average no. of summings of 2 primes/permutations for each of the 9,737 even numbers = 1,440,000 ÷ 9,737 = **147.89**

(7) <u>1,400 Consecutive Primes From 3 To 11,677</u>
(a) Even numbers (obtained by summing of 2 primes) = 6 to 23,354
(b) No. of even numbers = 11,675
(c) No. of primes = 1,400
(d) Average no. of even numbers "generated" by each of these 1,400 consecutive primes = 11,675 ÷ 1,400 = **8.34**
(e) No. of summings of 2 primes/permutations (3 + 3, 3 + 5, 3 + 7, 3 + 11, ... etc.) for these 1,400 primes = 1,400 x 1,400 = 1,960,000
(f) Average no. of summings of 2 primes/permutations for each of the 11,675 even numbers = 1,960,000 ÷ 11,675 = **167.88**

(8) <u>1,600 Consecutive Primes From 3 To 13,513</u>
(a) Even numbers (obtained by summing of 2 primes) = 6 to 27,026
(b) No. of even numbers = 13,511
(c) No. of primes = 1,600
(d) Average no. of even numbers "generated" by each of these 1,600 consecutive primes = 13,511 ÷ 1,600 = **8.44**
(e) No. of summings of 2 primes/permutations (3 + 3, 3 + 5, 3 + 7, 3 + 11, ... etc.) for these 1,600 primes = 1,600 x 1,600 = 2,560,000
(f) Average no. of summings of 2 primes/permutations for each of the 13,511 even numbers = 2,560,000 ÷ 13,511 = **189.48**

(9) <u>1,800 Consecutive Primes From 3 To 15,413</u>
(a) Even numbers (obtained by summing of 2 primes) = 6 to 30,826
(b) No. of even numbers = 15,411
(c) No. of primes = 1,800
(d) Average no. of even numbers "generated" by each of these 1,800 consecutive primes = 15,411 ÷ 1,800 = **8.56**
(e) No. of summings of 2 primes/permutations (3 + 3, 3 + 5, 3 + 7, 3 + 11, ... etc.) for these 1,800 primes = 1,800 x 1,800 = 3,240,000
(f) Average no. of summings of 2 primes/permutations for each of the 15,411

even numbers = 3,240,000 ÷ 15,411 = **210.24**

(10) <u>2,000 Consecutive Primes From 3 To 17,393</u>
   (a) Even numbers (obtained by summing of 2 primes) = 6 to 34,786
   (b) No. of even numbers = 17,391
   (c) No. of primes = 2,000
   (d) Average no. of even numbers "generated" by each of these 2,000 consecutive primes = 17,391 ÷ 2,000 = **8.70**
   (e) No. of summings of 2 primes/permutations (3 + 3, 3 + 5, 3 + 7, 3 + 11, ... etc.) for these 2,000 primes = 2,000 x 2,000 = 4,000,000
   (f) Average no. of summings of 2 primes/permutations for each of the 17,391 even numbers = 4,000,000 ÷ 17,391 = **230.00**

(11) <u>2,200 Consecutive Primes From 3 To 19,427</u>
   (a) Even numbers (obtained by summing of 2 primes) = 6 to 38,854
   (b) No. of even numbers = 19,425
   (c) No. of primes = 2,200
   (d) Average no. of even numbers "generated" by each of these 2,200 consecutive primes = 19,425 ÷ 2,200 = **8.83**
   (e) No. of summings of 2 primes/permutations (3 + 3, 3 + 5, 3 + 7, 3 + 11, ... etc.) for these 2,200 primes = 2,200 x 2,200 = 4,840,000
   (f) Average no. of summings of 2 primes/permutations for each of the 19,425 even numbers = 4,840,000 ÷ 19,425 = **249.16**

(12) <u>2,400 Consecutive Primes From 3 To 21,391</u>
   (a) Even numbers (obtained by summing of 2 primes) = 6 to 42,782
   (b) No. of even numbers = 21,389
   (c) No. of primes = 2,400
   (d) Average no. of even numbers "generated" by each of these 2,400 consecutive primes = 21,389 ÷ 2,400 = **8.91**
   (e) No. of summings of 2 primes/permutations (3 + 3, 3 + 5, 3 + 7, 3 + 11, ... etc.) for these 2,400 primes = 2,400 x 2,400 = 5,760,000
   (f) Average no. of summings of 2 primes/permutations for each of the 21,389 even numbers = 5,760,000 ÷ 21,389 = **269.30**

There would evidently be more and more profuse repetitions and overlaps of the even numbers "generated" by the primes the higher up the infinite list of prime numbers we go, which is significant. (For a better insight of this, refer to Appendix 1 and Appendix 2.)

We compare all the (d)s and (f)s in (1) to (12) above, which is as follows:-

(d) Average no. of even numbers "generated" by each of the consecutive primes in (1) to (12) above, as follows according to the listings (1) to (12):

   (1) **6.14**
   (2) **6.87**
   (3) **7.37**
   (4) **7.68**
   (5) **7.93**
   (6) **8.11**
   (7) **8.34**
   (8) **8.44**

(9) **8.56**
(10) **8.70**
(11) **8.83**
(12) **8.91**

(f) Average no. of summings of 2 primes/permutations for each of the even numbers in (1) to (12) above, as follows according to the listings (1) to (12):

(1) **32.60**
(2) **58.25**
(3) **81.47**
(4) **104.22**
(5) **126.18**
(6) **147.89**
(7) **167.88**
(8) **189.48**
(9) **210.24**
(10) **230.00**
(11) **249.16**
(12) **269.30**

The following is evident from the above information:-

(A): (d) Average no. of even numbers "generated" by each of the consecutive primes in the above 12 listings increases continually all the way from the list: (1) 200 Consecutive Primes From 3 To 1,229 to the list: (12) 2,400 Consecutive Primes From 3 To 21,391, from **6.14** even numbers per prime number in List (1) to **8.91** even numbers per prime number in List (12).

(B): (f) Average no. of summings of 2 primes/permutations for each of the even numbers in the above 12 listings increases continually all the way from the list: (1) 200 Consecutive Primes From 3 To 1,229 to the list: (12) 2,400 Consecutive Primes From 3 To 21,391, from **32.60** number of summings of 2 primes/permutations per even number in List (1) to **269.30** number of summings of 2 primes/permutations per even number in List (12).

Lemma. According to the principle of complete induction in set theory, if a set of natural numbers contains 1 and, for each n, it contains $n + 1$ whenever it contains all numbers less than $n + 1$, then it must contain every natural number, e.g., complete induction proves that every natural number is a product of primes.

By induction, we now deduce the following:

The larger the list of consecutive primes becomes, the greater would be the average number of even numbers "generated" by each of the primes in the list of consecutive primes (inferred from (A) above).

The larger the list of consecutive primes becomes, the greater would be the average number of summings of 2 primes/permutations for each of the even numbers in the infinite list of even numbers (inferred from (B) above).

Furthermore, the Goldbach conjecture had been tested and found to be correct for every even number up to 12 x 10^{17}, which is not a small list, by a distributed computer search carried out at the University of Aveiro, Portugal, in 2008.

As the primes and the even numbers are infinite, by the above lemma and all the above deductions and information, it could be inferred that the increases stated in (A) and (B) above, with the even numbers each being the sum of 2 primes, continue to infinity, *i.e., the Goldbach conjecture becomes stronger and stronger the higher up the infinite list of prime numbers/even numbers we go - all the way to infinity.*

Next, we resort to the argument by contradiction. The above deduction would be reversed if, e.g., the following takes place (which is the reversal of the above-mentioned information):

(A): (d) Average no. of even numbers "generated" by each of the consecutive primes in the above 12 listings decreases continually all the way from the list: (1) 200 Consecutive Primes From 3 To 1,229 to the list: (12) 2,400 Consecutive Primes From 3 To 21,391, from **8.91** even numbers per prime number in List (1) to **6.14** even numbers per prime number in List (12).

(B): (f) Average no. of summings of 2 primes/permutations for each of the even numbers in the above 12 listings decreases continually all the way from the list: (1) 200 Consecutive Primes From 3 To 1,229 to the list: (12) 2,400 Consecutive Primes From 3 To 21,391, from **269.30** number of summings of 2 primes/permutations per even number in List (1) to **32.60** number of summings of 2 primes/permutations per even number in List (12).

If this reversed state happens, the implication is that there would reach a point when there are no more batches of 2 prime numbers summing together to form even numbers, in which case the Goldbach conjecture would be false. Evidently this would happen when the prime numbers are finite. As the prime numbers are infinite (as Euclid had proved long ago) this would never happen.

Since the above information indicate otherwise, and, the prime numbers are infinite, we could accept the above induction/deduction and infer that the Goldbach conjecture is not false, i.e., the Goldbach conjecture is true, and, every even number (except 2) is the sum of 2 prime numbers. This possibly concludes the argument by contradiction.

We take a look at the following example to see how effectively the primes "generate" new even numbers in accordance with the Goldbach conjecture:-

Density Of New Even Numbers "Generated" (*See Appendix 1 For Example Of Computation Method*)

(a) Set Of Integers, 51 To 100, With 10 Primes Within It = 5 New Even Nos. Per Prime No.
(No. Of New Even Nos. "Generated" = 50. No. Of Primes = 10.)

(b) Set Of Integers, 101 To 150, With 10 Primes Within It = 5.2 New Even Nos. Per Prime No.
(No. Of New Even Nos. "Generated" = 52. No. Of Primes = 10.)

(c) Set Of Integers, 151 To 200, With 11 Primes Within It = **4.55 New Even Nos. Per Prime No.**
(No. Of New Even Nos. "Generated" = 50. No. Of Primes = 11.)

(d) Set Of Integers, 201 To 250, With 7 Primes Within It = 6 New Even Nos. Per Prime No.
(No. Of New Even Nos. "Generated" = 42. No. Of Primes = 7.)

(e) Set Of Integers, 251 To 300, With 9 Primes Within It = 5.78 New Even Nos. Per Prime No.
(No. Of New Even Nos. "Generated" = 52. No. Of Primes = 9.)

(f) Set Of Integers, 301 To 350, With 8 Primes Within It = 7 New Even Nos. Per Prime No.
(No. Of New Even Nos. "Generated" = 56. No. Of Primes = 8.)

(g) Set Of Integers, 351 To 400, With 8 Primes Within It = 6 New Even Nos. Per Prime No.
(No. Of New Even Nos. "Generated" = 48. No. Of Primes = 8.)

(h) Set Of Integers, 401 To 450, With 9 Primes Within It = 5.78 New Even Nos. Per Prime No.
(No. Of New Even Nos. "Generated" = 52. No. Of Primes = 9.)

(i) Set Of Integers, 451 To 500, With 8 Primes Within It = 6.25 New Even Nos. Per Prime No.
(No. Of New Even Nos. "Generated" = 50. No. Of Primes = 8.)

(j) Set Of Integers, 501 To 550, With 6 Primes Within It = 8 New Even Nos. Per Prime No.
(No. Of New Even Nos. "Generated" = 48. No. Of Primes = 6.)

(k) Set Of Integers, 551 To 600, With 8 Primes Within It = 6.5 New Even Nos. Per Prime No.
(No. Of New Even Nos. "Generated" = 52. No. Of Primes = 8.)

(l) Set Of Integers, 601 To 650, With 9 Primes Within It = 5.33 New Even Nos. Per Prime No.
(No. Of New Even Nos. "Generated" = 48. No. Of Primes = 9.)

(m) Set Of Integers, 651 To 700, With 7 Primes Within It = 6.29 New Even Nos. Per Prime No.
(No. Of New Even Nos. "Generated" = 44. No. Of Primes = 7.)

(n) Set Of Integers, 701 To 750, With 7 Primes Within It = 7.43 New Even Nos. Per Prime No.
(No. Of New Even Nos. "Generated" = 52. No. Of Primes = 7.)

(o) Set Of Integers, 751 To 800, With 7 Primes Within It = 7.71 New Even Nos. Per Prime No.
(No. Of New Even Nos. "Generated" = 54. No. Of Primes = 7.)

(p) Set Of Integers, 801 To 850, With 7 Primes Within It = 6 New Even Nos. Per Prime No.
(No. Of New Even Nos. "Generated" = 42. No. Of Primes = 7.)

(q) Set Of Integers, 851 To 900, With 8 Primes Within It = 6 New Even Nos. Per Prime No.
(No. Of New Even Nos. "Generated" = 48. No. Of Primes = 8.)

(r) Set Of Integers, 901 To 950, With 7 Primes Within It = 8.57 New Even Nos. Per Prime No.
(No. Of New Even Nos. "Generated" = 60. No. Of Primes = 7.)

(s) Set Of Integers, 951 To 1,000, With 7 Primes Within It = 7.14 New Even Nos. Per Prime No.
(No. Of New Even Nos. "Generated" = 50. No. Of Primes = 7.)

(t) Set Of Integers, 1,001 To 1,050, With 8 Primes Within It = 6.5 New Even Nos. Prime No.
(No. Of New Even Nos. "Generated" = 52. No. Of Primes = 8.)

(u) Set Of Integers, 1,051 To 1,100, With 8 Primes Within It = 6 New Even Nos. Per Prime No.

(No. Of New Even Nos. "Generated" = 48. No. Of Primes = 8.)

(v) Set Of Integers, 1,101 To 1,150, With 5 Primes Within It = 6.4 New Even Nos. Per Prime No.
(No. Of New Even Nos. "Generated" = 32. No. Of Primes = 5.)

(w) Set Of Integers, 1,151 To 1,200, With 7 Primes Within It = **9.14 New Even Nos. Per Prime No.**
(No. Of New Even Nos. "Generated" = 64. No. Of Primes = 7.)

(x) Set Of Integers, 1,201 To 1,250, With 8 Primes Within It = 7 New Even Nos. Per Prime No.
(No. Of New Even Nos. "Generated" = 56. No. Of Primes = 8.)

Average Density For The Above 24 Items ((a) To (x)) = 155.54 ÷ 24 = 6.48 New Even Nos. Per Prime No.

Maximum Density = 9.14 New Even Nos. Per Prime No. (No. Of New Even Nos. "Generated" = 64. No. Of Primes = 7.)

Minimum Density = 4.55 New Even Nos. Per Prime No. (No. Of New Even Nos. "Generated" = 50. No. Of Primes = 11.)

Such a "profuse generation" of "regular batches" of even numbers by the prime numbers is significant and lends further support to the possibly validity of the Goldbach conjecture.

There is further argument which is obtainable by analysing a number of even numbers, e.g., we could split a group of 240 even consecutive numbers, from 4 to 482, into 8 equal batches (30 even numbers per batch) and analyse the batches; this would corroborate the fact that the infinite quantity of primes would "generate" a regular, continuous (*without breaks or gaps*) and infinite list of even numbers. The density of distribution or prime additions/combinations per even number evidently become greater and greater the higher up the infinite list of the even numbers we go, i.e., *the Goldbach conjecture evidently becomes stronger and stronger the higher up the infinite list of the even numbers we go*. This pattern is significant and is discernable in the following example:-

(1) <u>1 st. Batch Of 30 Even Numbers (4 To 62)</u> (*See Appendix 2 For Example Of Computation Method*)
    a) Maximum No. Of Prime Additions/Combinations Per Even Number = **5**
    b) Minimum No. Of Prime Additions/Combinations Per Even Number = **1**
    c) Density Of Distribution = Average Prime Additions/Combinations Per Even Number = **2.77** Prime Additions/Combinations Per Even Number

(2) <u>2 nd. Batch Of 30 Even Numbers (64 To 122)</u>

    a) Maximum No. Of Prime Additions/Combinations Per Even Number = 14
    b) Minimum No. Of Prime Additions/Combinations Per Even Number = 2
    c) Density Of Distribution = Average Prime Additions/Combinations Per Even Number = **6.1** Prime Additions/Combinations Per Even Number
    d) Percentage Increase In Density Of Distribution = (6.1 - 2.77) ÷ 2.77 x 100% = 120.22%

(3) <u>3 rd. Batch Of 30 Even Numbers (124 To 182)</u>

    a) Maximum No. Of Prime Additions/Combinations Per Even Number = 16
    b) Minimum No. Of Prime Additions/Combinations Per Even Number = 4

c) Density Of Distribution = Average Prime Additions/Combinations Per Even Number = **9.07** Prime Additions/Combinations Per Even Number

d) Percentage Increase In Density Of Distribution = (9.07 - 6.1) ÷ 6.1 x 100% = 48.69%

(4) <u>4 th. Batch Of 30 Even Numbers (184 To 242)</u>

a) Maximum No. Of Prime Additions/Combinations Per Even Number = 22
b) Minimum No. Of Prime Additions/Combinations Per Even Number = 5
c) Density Of Distribution = Average Prime Additions/Combinations Per Even Number = **10.53** Prime Additions/Combinations Per Even Number
d) Percentage Increase In Density Of Distribution = (10.53 - 9.07) ÷ 9.07 x 100% = 16.1%

(5) <u>5 th. Batch Of 30 Even Numbers (244 To 302)</u>

a) Maximum No. Of Prime Additions/Combinations Per Even Number = 21
b) Minimum No. Of Prime Additions/Combinations Per Even Number = 7
c) Density Of Distribution = Average Prime Additions/Combinations Per Even Number = **12.37** Prime Additions/Combinations Per Even Number
d) Percentage Increase In Density Of Distribution = (12.37 - 10.53) ÷ 10.53 x 100% = 17.47%

(6) <u>6 th. Batch Of 30 Even Numbers (304 To 362)</u>

a) Maximum No. Of Prime Additions/Combinations Per Even Number = 27
b) Minimum No. Of Prime Additions/Combinations Per Even Number = 7
c) Density Of Distribution = Average Prime Additions/Combinations Per Even Number = **13.77** Prime Additions/Combinations Per Even Number
d) Percentage Increase In Density Of Distribution = (13.77 - 12.37) ÷ 12.37 x 100% = 11.32%

(7) <u>7 th. Batch Of 30 Even Numbers (364 To 422)</u>

a) Maximum No. Of Prime Additions/Combinations Per Even Number = 30
b) Minimum No. Of Prime Additions/Combinations Per Even Number = 7
c) Density Of Distribution = Average Prime Additions/Combinations Per Even Number = **15.23** Prime Additions/Combinations Per Even Number
d) Percentage Increase In Density Of Distribution = (15.23 - 13.77) ÷ 13.77 x 100% = 10.6%

(8) <u>8 th. Batch Of 30 Even Numbers (424 To 482)</u>

a) Maximum No. Of Prime Additions/Combinations Per Even Number = **30**
b) Minimum No. Of Prime Additions/Combinations Per Even Number = **9**
c) Density Of Distribution = Average Prime Additions/Combinations Per Even Number = **16.93** Prime Additions/Combinations Per Even Number
d) Percentage Increase In Density Of Distribution = (16.93 - 15.23) ÷ 15.23 x 100% = 11.16%

The Density Of Distribution is expected to increase to infinity, though the Percentage Increase In Density Of Distribution is expected to thin out towards infinity - it could be seen above to increase from 2.77 prime additions/combinations per even number for batch of even numbers, 4 to 62, all the way up to 16.93 prime additions/combinations per even number for batch of even numbers, 424 to 482. This is nevertheless significant evidence that lends support to the possibly validity of the Goldbach conjecture. Also, the Maximum No. Of Prime Additions/Combinations Per Even Number and the Minimum No. Of Prime Additions/Combinations Per Even Number could be seen to range from 5 and 1 respectively for batch of even numbers, 4 to 62, to 30 and 9 respectively for batch of even numbers, 424 to 482. This trend of "upward increase" of the (maximum and minimum) numbers of prime additions/combinations for each even number implies that at some points toward infinity the numbers of prime additions/combinations for each even number could be thousands, millions, billions, trillions, and more, if only we have the computing power to compute/check such prime additions/combinations (this again indicates that *the Goldbach conjecture becomes evidently stronger and stronger the higher up the infinite list of the even numbers we go*). This is significant too and is also evidence that lends support to the possibly validity of the Goldbach conjecture. By the infinitude of the primes (vide Euclid's theorem) and even numbers, these "patterns", as described here, would possibly be there all the way to infinity, which would be in accordance with the Goldbach conjecture.

The following evidence would possibly further affirm the possibly validity of the Goldbach conjecture:-

1) 10 consecutive primes, commencing from the odd prime 3, would give rise to 10 x 10, or, 100 sums of 2 primes/partitions/permutations, but less than 100 different even numbers, with many repetitions/overlaps (e.g., for these first 10 consecutive primes 3, 5, 7, 11, 13, 17, 19, 23, 29 & 31, 10 = 3 + 7 = 5 + 5 (2 partitions/permutations), 22 = 3 + 19 = 5 + 17 = 11 + 11 (3 partitions/permutations), &, 34 = 3 + 31 = 5 + 29 = 11 + 23 = 17 + 17 (4 partitions/permutations)).
2) 20 consecutive primes, commencing from the odd prime 3, (increase of **100%** in no. of consecutive primes compared to (1) above) would give rise to 20 x 20, or, 400 sums of 2 primes/partitions/permutations (increase of **300%** in no. of sums of 2 primes/partitions/permutations compared to (1) above), but less than 400 different even numbers, with many repetitions/overlaps.
3) 30 consecutive primes, commencing from the odd prime 3, (increase of **200%** in no. of consecutive primes compared to (1) above) would give rise to 30 x 30, or, 900 sums of 2 primes/partitions/permutations (increase of **800%** in no. of sums of 2 primes/partitions/permutations compared to (1) above), but less than 900 different even numbers, with many repetitions/overlaps.
4) 40 consecutive primes, commencing from the odd prime 3, (increase of **300%** in no. of consecutive primes compared to (1) above) would give rise to 40 x 40, or, 1,600 sums of 2 primes/partitions/permutations (increase of **1,500%** in no. of sums of 2 primes/partitions/permutations compared to (1) above), but less than 1,600 different even numbers, with many repetitions/overlaps.
5) 50 consecutive primes, commencing from the odd prime 3, (increase of **400%** in no. of consecutive primes compared to (1) above) would give rise to 50 x 50, or, 2,500 sums of 2 primes/partitions/permutations (increase of **2,400%** in no. of sums of 2 primes/partitions/permutations compared to (1) above), but less than 2,500 different even numbers, with many repetitions/overlaps.
6) 60 consecutive primes, commencing from the odd prime 3, (increase of **500%** in no. of consecutive primes compared to (1) above) would give rise to 60 x 60, or, 3,600 sums of 2 primes/partitions/permutations (increase of **3,500%** in no. of sums of 2 primes/partitions/permutations compared to (1) above), but less than 3,600 different even numbers, with many repetitions/overlaps.

7) 70 consecutive primes, commencing from the odd prime 3, (increase of **600%** in no. of consecutive primes compared to (1) above) would give rise to 70 x 70, or, 4,900 sums of 2 primes/partitions/permutations (increase of **4,800%** in no. of sums of 2 primes/partitions/permutations compared to (1) above), but less than 4,900 different even numbers, with many repetitions/overlaps.
8) 80 consecutive primes, commencing from the odd prime 3, (increase of **700%** in no. of consecutive primes compared to (1) above) would give rise to 80 x 80, or, 6,400 sums of 2 primes/partitions/permutations (increase of **6,300%** in no. of sums of 2 primes/partitions/permutations compared to (1) above), but less than 6,400 different even numbers, with many repetitions/overlaps.
9) 90 consecutive primes, commencing from the odd prime 3, (increase of **800%** in no. of consecutive primes compared to (1) above) would give rise to 90 x 90, or, 8,100 sums of 2 primes/partitions/permutations (increase of **8,000%** in no. of sums of 2 primes/partitions/permutations compared to (1) above), but less than 8,100 different even numbers, with many repetitions/overlaps.
10) 100 consecutive primes, commencing from the odd prime 3, (increase of **900%** in no. of consecutive primes compared to (1) above) would give rise to 100 x 100, or, 10,000 sums of 2 primes/partitions/permutations (increase of **9,900%** in no. of sums of 2 primes/partitions/permutations compared to (1) above), but less than 10,000 different even numbers, with many repetitions/overlaps.

$$\vdots$$

The following is evident from the above:-

1) The 1st. marginal increase of **100%** in no. of consecutive primes (increase of 200% - increase of 100%) results in marginal increase of **500%** in no. of sums of 2 primes/partitions/permutations (increase of 800% - increase of 300%).
2) The 2nd. marginal increase of **100%** in no. of consecutive primes (increase of 300% - increase of 200%) results in marginal increase of **700%** in no. of sums of 2 primes/partitions/permutations (increase of 1,500% - increase of 800%).
3) The 3rd. marginal increase of **100%** in no. of consecutive primes (increase of 400% - increase of 300%) results in marginal increase of **900%** in no. of sums of 2 primes/partitions/permutations (increase of 2,400% - increase of 1,500%).
4) The 4th. marginal increase of **100%** in no. of consecutive primes (increase of 500% - increase of 400%) results in marginal increase of **1,100%** in no. of sums of 2 primes/partitions/permutations (increase of 3,500% - increase of 2,400%).
5) The 5th. marginal increase of **100%** in no. of consecutive primes (increase of 600% - increase of 500%) results in marginal increase of **1,300%** in no. of sums of 2 primes/partitions/permutations (increase of 4,800% - increase of 3,500%).
6) The 6th. marginal increase of **100%** in no. of consecutive primes (increase of 700% - increase of 600%) results in marginal increase of **1,500%** in no. of sums of 2 primes/partitions/permutations (increase of 6,300% - increase of 4,800%).
7) The 7th. marginal increase of **100%** in no. of consecutive primes (increase of 800% - increase of 700%) results in marginal increase of **1,700%** in no. of sums of 2 primes/partitions/permutations (increase of 8,000% - increase of 6,300%).
8) The 8th. marginal increase of **100%** in no. of consecutive primes (increase of 900% - increase of 800%) results in marginal increase of **1,900%** in no. of sums of 2 primes/partitions/permutations (increase of 9,900% - increase of 8,000%).

(1) to (8) above show that while the marginal increase in no. of consecutive primes remains constant at 100% from (1) to (8), the marginal increase in no. of sums of 2 primes/partitions/permutations goes up progressively from 500% in (1) to 1,900% in (8). It is evident here that the higher up the infinite list of primes we go, the more "overwhelming" or dense the (one-to-one) combinations of primes (i.e., sums of 2 primes, in the formation of even numbers) would become, the number of permutations of the combinations of primes tending towards infinity (with the infinity of the prime numbers). *In other words, the Goldbach conjecture becomes stronger and stronger the higher up the infinite list of prime numbers/even numbers we go.* The infinitude of the prime numbers (vide Euclid's theorem) and even numbers would hence imply the possibly validity of the Goldbach conjecture.

The prime number theorem, which had been proven independently by Hadamard and de la Vallee Poussin in 1896, states that the limit of the quotient of the 2 functions $\pi(n)$ and $n/\log n$ as n, which is a positive real number, approaches infinity is 1, which is expressed by the formula:

$$\lim_{n \to \infty} \pi(n)/(n/\log n) = 1$$

where $\pi(n)$ is approximately equal to $(n/\log n)$

The function $\pi(n)$ represents the number of primes not greater than the number n. This function measures the distribution of the prime numbers. With it, we compute the ratio $n/\pi(n)$ which says what fraction of the numbers up to a given point are primes. (It is actually the reciprocal of this fraction.) The following is the result of a computation:-

n	$\pi(n)$	$n/\pi(n)$
10	4 (a)	2.5
100	25 (b)	4.0
1,000	168 (c)	6.0
10,000	1,229 (d)	8.1
100,000	9,592 (e)	10.4
1,000,000	78,498 (f)	12.7
10,000,000	664,579 (g)	15.0
100,000,000	5,761,455 (h)	17.4
1,000,000,000	50,847,534 (i)	19.7
10,000,000,000	455,052,512 (j)	22.0

It is noticeable that as one moves from 1 power of 10 to the next, the ratio $n/\pi(n)$ increases by about 2.3, e.g., 22.0 - 19.7 = 2.3. As $\log_e 10 = 2.30258 \ldots$, we may thus regard $\pi(n)$ as approximately equal to $n/\log n$.

We have the following partitions with the primes described in the "$\pi(n)$" column above:-

1) With (a) above, we have the following "prime + prime = even number" combinations:

   a) prime a + prime a: 4 x 4 "prime + prime" combinations
   b) prime a + prime b: 4 x 25 "prime + prime" combinations

c) prime a + prime c: 4 x 168 "prime + prime" combinations
d) prime a + prime d: 4 x 1,229 "prime + prime" combinations
e) prime a + prime e: 4 x 9,592 "prime + prime" combinations
f) prime a + prime f: 4 x 78,498 "prime + prime" combinations
g) prime a + prime g: 4 x 664,579 "prime + prime" combinations
h) prime a + prime h: 4 x 5,761,455 "prime + prime" combinations
i) prime a + prime i: 4 x 50,847,534 "prime + prime" combinations
j) prime a + prime j: 4 x 455,052,512 "prime + prime" combinations

For example, for (j) above, a prime described in (a) in the "$\pi(n)$" column above plus a prime described in (j) in the "$\pi(n)$" column above give an even number, and there are 4 x 455,052,512 such "prime + prime = even number" combinations.

2) With (b) above, we have the following "prime + prime = even number" combinations:

a) prime b + prime a: 25 x 4 "prime + prime" combinations
b) prime b + prime b: 25 x 25 "prime + prime" combinations
c) prime b + prime c: 25 x 168 "prime + prime" combinations
d) prime b + prime d: 25 x 1,229 "prime + prime" combinations
e) prime b + prime e: 25 x 9,592 "prime + prime" combinations
f) prime b + prime f: 25 x 78,498 "prime + prime" combinations
g) prime b + prime g: 25 x 664,579 "prime + prime" combinations
h) prime b + prime h: 25 x 5,761,455 "prime + prime" combinations
i) prime b + prime i: 25 x 50,847,534 "prime + prime" combinations
j) prime b + prime j: 25 x 455,052,512 "prime + prime" combinations

3) With (c) above, we have the following "prime + prime = even number" combinations:

a) prime c + prime a: 168 x 4 "prime + prime" combinations
b) prime c + prime b: 168 x 25 "prime + prime" combinations
c) prime c + prime c: 168 x 168 "prime + prime" combinations
d) prime c + prime d: 168 x 1,229 "prime + prime" combinations
e) prime c + prime e: 168 x 9,592 "prime + prime" combinations
f) prime c + prime f: 168 x 78,498 "prime + prime" combinations
g) prime c + prime g: 168 x 664,579 "prime + prime" combinations
h) prime c + prime h: 168 x 5,761,455 "prime + prime" combinations
i) prime c + prime i: 168 x 50,847,534 "prime + prime" combinations
j) prime c + prime j: 168 x 455,052,512 "prime + prime" combinations

4) With (d) above, we have the following "prime + prime = even number" combinations:

a) prime d + prime a: 1,229 x 4 "prime + prime" combinations
b) prime d + prime b: 1,229 x 25 "prime + prime" combinations
c) prime d + prime c: 1,229 x 168 "prime + prime" combinations
d) prime d + prime d: 1,229 x 1,229 "prime + prime" combinations
e) prime d + prime e: 1,229 x 9,592 "prime + prime" combinations
f) prime d + prime f: 1,229 x 78,498 "prime + prime" combinations
g) prime d + prime g: 1,229 x 664,579 "prime + prime" combinations

h) prime d + prime h: 1,229 x 5,761,455 "prime + prime" combinations
   i) prime d + prime i: 1,229 x 50,847,534 "prime + prime" combinations
   j) prime d + prime j: 1,229 x 455,052,512 "prime + prime" combinations

5) With (e) above, we have the following "prime + prime = even number" combinations:

   a) prime e + prime a: 9,592 x 4 "prime + prime" combinations
   b) prime e + prime b: 9,592 x 25 "prime + prime" combinations
   c) prime e + prime c: 9,592 x 168 "prime + prime" combinations
   d) prime e + prime d: 9,592 x 1,229 "prime + prime" combinations
   e) prime e + prime e: 9,592 x 9,592 "prime + prime" combinations
   f) prime e + prime f: 9,592 x 78,498 "prime + prime" combinations
   g) prime e + prime g: 9,592 x 664,579 "prime + prime" combinations
   h) prime e + prime h: 9,592 x 5,761,455 "prime + prime" combinations
   i) prime e + prime i: 9,592 x 50,847,534 "prime + prime" combinations
   j) prime e + prime j: 9,592 x 455,052,512 "prime + prime" combinations

6) With (f) above, we have the following "prime + prime = even number" combinations:

   a) prime f + prime a: 78,498 x 4 "prime + prime" combinations
   b) prime f + prime b: 78,498 x 25 "prime + prime" combinations
   c) prime f + prime c: 78,498 x 168 "prime + prime" combinations
   d) prime f + prime d: 78,498 x 1,229 "prime + prime" combinations
   e) prime f + prime e: 78,498 x 9,592 "prime + prime" combinations
   f) prime f + prime f: 78,498 x 78,498 "prime + prime" combinations
   g) prime f + prime g: 78,498 x 664,579 "prime + prime" combinations
   h) prime f + prime h: 78,498 x 5,761,455 "prime + prime" combinations
   i) prime f + prime i: 78,498 x 50,847,534 "prime + prime" combinations
   j) prime f + prime j: 78,498 x 455,052,512 "prime + prime" combinations

7) With (g) above, we have the following "prime + prime = even number" combinations:

   a) prime g + prime a: 664,579 x 4 "prime + prime" combinations
   b) prime g + prime b: 664,579 x 25 "prime + prime" combinations
   c) prime g + prime c: 664,579 x 168 "prime + prime" combinations
   d) prime g + prime d: 664,579 x 1,229 "prime + prime" combinations
   e) prime g + prime e: 664,579 x 9,592 "prime + prime" combinations
   f) prime g + prime f: 664,579 x 78,498 "prime + prime" combinations
   g) prime g + prime g: 664,579 x 664,579 "prime + prime" combinations
   h) prime g + prime h: 664,579 x 5,761,455 "prime + prime" combinations
   i) prime g + prime i: 664,579 x 50,847,534 "prime + prime" combinations
   j) prime g + prime j: 664,579 x 455,052,512 "prime + prime" combinations

8) With (h) above, we have the following "prime + prime = even number" combinations:

   a) prime h + prime a: 5,761,455 x 4 "prime + prime" combinations
   b) prime h + prime b: 5,761,455 x 25 "prime + prime" combinations

c) prime h + prime c: 5,761,455 x 168 "prime + prime" combinations
d) prime h + prime d: 5,761,455 x 1,229 "prime + prime" combinations
e) prime h + prime e: 5,761,455 x 9,592 "prime + prime" combinations
f) prime h + prime f: 5,761,455 x 78,498 "prime + prime" combinations
g) prime h + prime g: 5,761,455 x 664,579 "prime + prime" combinations
h) prime h + prime h: 5,761,455 x 5,761,455 "prime + prime" combinations
i) prime h + prime i: 5,761,455 x 50,847,534 "prime + prime" combinations
j) prime h + prime j: 5,761,455 x 455,052,512 "prime + prime" combinations

9) With (i) above, we have the following "prime + prime = even number" combinations:

a) prime i + prime a: 50,847,534 x 4 "prime + prime" combinations
b) prime i + prime b: 50,847,534 x 25 "prime + prime" combinations
c) prime i + prime c: 50,847,534 x 168 "prime + prime" combinations
d) prime i + prime d: 50,847,534 x 1,229 "prime + prime" combinations
e) prime i + prime e: 50,847,534 x 9,592 "prime + prime" combinations
f) prime i + prime f: 50,847,534 x 78,498 "prime + prime" combinations
g) prime i + prime g: 50,847,534 x 664,579 "prime + prime" combinations
h) prime i + prime h: 50,847,534 x 5,761,455 "prime + prime" combinations
i) prime i + prime i: 50,847,534 x 50,847,534 "prime + prime" combinations
j) prime i + prime j: 50,847,534 x 455,052,512 "prime + prime" combinations

10) With (j) above, we have the following "prime + prime = even number" combinations:

a) prime j + prime a: 455,052,512 x 4 "prime + prime" combinations
b) prime j + prime b: 455,052,512 x 25 "prime + prime" combinations
c) prime j + prime c: 455,052,512 x 168 "prime + prime" combinations
d) prime j + prime d: 455,052,512 x 1,229 "prime + prime" combinations
e) prime j + prime e: 455,052,512 x 9,592 "prime + prime" combinations
f) prime j + prime f: 455,052,512 x 78,498 "prime + prime" combinations
g) prime j + prime g: 455,052,512 x 664,579 "prime + prime" combinations
h) prime j + prime h: 455,052,512 x 5,761,455 "prime + prime" combinations
i) prime j + prime i: 455,052,512 x 50,847,534 "prime + prime" combinations
j) prime j + prime j: 455,052,512 x 455,052,512 "prime + prime" combinations

.
.
.

The above partitions/"prime + prime = even number" combinations are evidently progressively more "overwhelming", dense (refer to Figure 1 below), and repetitive (overlapping). *That is, the Goldbach conjecture becomes evidently progressively stronger and stronger towards infinity, which corroborates the earlier observation/induction.* It is not surprising that computer searches completed in 2000 had verified that all even numbers up to 400 trillion (4×10^{14}), which is not a small list, are sums of 2 primes, while in 2008, a distributed computer search ran by Tomas Oliveira e Silva, a researcher at the University of Aveiro, Portugal, had further verified the Goldbach conjecture up to 12×10^{17}, which is a long, impressive list.

Though the distribution of primes evidently becomes progressively less and less dense, e.g., ranging from 40% of primes within the first 10 integers to 4.55% of primes within the first 10,000,000,000 integers, the density of partitions/"prime + prime = even number" combinations evidently becomes progressively greater and greater as is shown below:-

1) For the 1st. **10-fold** increase in no. of integers (100 integers ÷ 10 integers), the no. of partitions/"prime + prime = even number" combinations increases **39.06 times** ([25 x 25 partitions] ÷ [4 x 4 partitions]).
2) For the 2nd. **10-fold** increase in no. of integers (1,000 integers ÷ 100 integers), the no. of partitions/"prime + prime = even number" combinations increases **45.16 times** ([168 x 168 partitions] ÷ [25 x 25 partitions]).
3) For the 3rd. **10-fold** increase in no. of integers (10,000 integers ÷ 1,000 integers), the no. of partitions/"prime + prime = even number" combinations increases **53.52 times** ([1,229 x 1,229 partitions] ÷ [168 x 168 partitions]).
4) For the 4th. **10-fold** increase in no. of integers (100,000 integers ÷ 10,000 integers), the no. of partitions/"prime + prime = even number" combinations increases **60.91 times** ([9,592 x 9,592 partitions] ÷ [1,229 x 1,229 partitions]).
5) For the 5th. **10-fold** increase in no. of integers (1,000,000 integers ÷ 100,000 integers), the no. of partitions/"prime + prime = even number" combinations increases **66.97 times** ([78,498 x 78,498 partitions] ÷ [9,592 x 9,592 partitions]).
6) For the 6th. **10-fold** increase in no. of integers (10,000,000 integers ÷ 1,000,000 integers), the no. of partitions/"prime + prime = even number" combinations increases **71.68 times** ([664,579 x 664,579 partitions] ÷ [78,498 x 78,498 partitions]).
7) For the 7th. **10-fold** increase in no. of integers (100,000,000 integers ÷ 10,000,000 integers), the no. of partitions/"prime + prime = even number" combinations increases **75.16 times** ([5,761,455 x 5,761,455 partitions] ÷ [664,579 x 664,579 partitions]).
8) For the 8th. **10-fold** increase in no. of integers (1,000,000,000 integers ÷ 100,000,000 integers), the no. of partitions/"prime + prime = even number" combinations increases **77.89 times** ([50,847,534 x 50,847,534 partitions] ÷ [5,761,455 x 5,761,455 partitions]).
9) For the 9th. **10-fold** increase in no. of integers (10,000,000,000 integers ÷ 1,000,000,000 integers), the no. of partitions/"prime + prime = even number" combinations increases **80.09 times** ([455,052,512 x 455,052,512 partitions] ÷ [50,847,534 x 50,847,534 partitions]).

<u>Figure 1</u>

The infinitude of the primes, as per Euclid's theorem, together with the infinitude of the even numbers, however imply that the above partitions/"prime + prime = even number" combinations would become increasingly more "overwhelming", dense, and repetitive (overlapping) towards infinity (*the Goldbach conjecture becoming evidently stronger and stronger the higher up the infinite list of prime numbers/even numbers we go*), hence "ensuring" the continuity (without any breaks or gaps) of the even numbers found, and would be so all the way to infinity, thus implying that possibly every even number after 2 is the sum of 2 primes. (For a better insight of how the above partitions/"prime + prime = even number" combinations would become increasingly more "overwhelming", dense, and repetitive (overlapping) towards infinity, refer to Appendix 1 and Appendix 2.)

The partitions/"prime + prime = even number" combinations, as had been conjectured by Goldbach, are evidently effusive, or in great abundance, in their occurrences, as is shown above and in the appendices below. This has important consequence. For instance, in Appendix 2, the number of partitions/"prime + prime = even number" combinations for each of the 30 even numbers (424 to 482) ranges from the minimum 9 (for the even numbers 428 and 458) to the maximum 30 (for the even numbers 462 and 480), giving an average of 16.93 partitions/"prime + prime = even number" combinations per even number. This is significant and is in stark contrast to the results of the Fundamental Theorem of Arithmetic or Unique Factorization Theorem, which states that there is only 1 possible combination of primes which will multiply together to produce any particular number, e.g., the only combination of primes which will produce the number 2,079 is as follows:-

$$3 \times 3 \times 3 \times 7 \times 11 \text{ (only)}$$

In the same manner, the following numbers are also uniquely factorized:-

$$63 = 3 \times 3 \times 7 \text{ (only)}$$
$$153 = 3 \times 3 \times 17 \text{ (only)}$$
$$1{,}021{,}020 = 2 \times 2 \times 3 \times 5 \times 7 \times 11 \times 13 \times 17 \text{ (only)}$$

In other words, every positive whole number can be broken up into prime factors, and, this can happen in only 1 way. In contrast, every even number is the sum of 2 primes in more than 1 way, e.g., 30 ways (i.e., 30 possible partitions) in the cases of the even numbers 462 and 480 as is described above. As is stated above, this is significant. This effusiveness or abundance of partitions/"prime + prime = even number" combinations somehow implies that the continuity (without any breaks or gaps) of the even numbers as sums of 2 primes (which are "generated" through the various additions of 2 primes) is "ensured", i.e., the possible breaks in the continuity of the even numbers as sums of 2 primes (wherein some even numbers in-between can never be sums of 2 primes, as are shown in the example in Figure 2 below where there are 4 breaks in the continuity of the even numbers x_1 to x_{12}, where the 4 even numbers x_4, x_5, x_8 & x_{10} can never be sums of 2 primes), which implies the falsity of the Goldbach conjecture, are somehow possibly "prevented from happening" by this effusiveness or abundance:-

x below represents, say, an extremely large even number. p below represents a prime. c below represents a composite number or non-prime whence the Goldbach conjecture would be false (i.e., not every even number is the sum of 2 primes as the composite numbers or non-primes would be the exceptions).

$$\vdots$$
$$x_1 = p_1 + p_2$$
$$x_2 = p_3 + p_4$$
$$x_3 = p_5 + p_6$$
$$x_4 = c_1 + c_2 \text{ (break)}$$
$$x_5 = p_7 + c_3 \text{ (break)}$$
$$x_6 = p_8 + p_9$$
$$x_7 = p_{10} + p_{11}$$
$$x_8 = c_4 + p_{12} \text{ (break)}$$
$$x_9 = p_{13} + p_{14}$$
$$x_{10} = c_5 + c_6 \text{ (break)}$$
$$x_{11} = p_{15} + p_{16}$$
$$x_{12} = p_{17} + p_{18}$$
$$\vdots$$

<u>Figure 2</u>

There appears to be some deep meaning in the ease and effusiveness with which the partitions/"prime + prime = even number" combinations or sums of 2 primes show up, as is shown in this chapter. If every even number which is the sum of 2 primes is the sum of 2 primes in only 1 way (a la the results of the Fundamental Theorem of Arithmetic or Unique Factorization Theorem described above), there could be possible breaks in the continuity of the even numbers as sums of 2 primes (as are shown in Figure 2 above), in other words, there could be some reason to doubt the validity of the Goldbach conjecture. But, on the contrary, the sums of 2 primes are evidently much effusive; they are evidently a defining characteristic of the even numbers. Under such a circumstance, it would be difficult to doubt the validity of the Goldbach conjecture.

In elaborating further on the above point, we take a look at the following:-

<u>No. Of Old/Repeated (Also Appeared Earlier) Even Numbers/Overlaps "Generated" (By The Additions/Combinations Of Two Primes), For Integers 1 To 1,250</u> (*See Appendix 1 For Example Of Computation Method*)

(a) Set Of Integers, 1 To 50, With 14 Primes Within It = Not Applicable
(aa) Percentage Increase In Repetition = Not Applicable

(b) Set Of Integers, 51 To 100, With 10 Primes Within It = **20** Repeated Even Nos.
(bb) Percentage Increase In Repetition = Not Applicable

(c) Set Of Integers, 101 To 150, With 10 Primes Within It = 46 Repeated Even Nos.
(cc) Percentage Increase In Repetition = (46 - 20) ÷ 20 x 100% = **130%**

(d) Set Of Integers, 151 To 200, With 11 Primes Within It = 73 Repeated Even Nos.
(dd) Percentage Increase In Repetition = (73 - 46) ÷ 46 x 100% = 58.7%

(e) Set Of Integers, 201 To 250, With 7 Primes Within It = 93 Repeated Even Nos.
(ee) Percentage Increase In Repetition = (93 - 73) ÷ 73 x 100% = 27.4%

(f) Set Of Integers, 251 To 300, With 9 Primes Within It = 115 Repeated Even Nos.
(ff) Percentage Increase In Repetition = (115 - 93) ÷ 93 x 100% = 23.66%

(g) Set Of Integers, 301 To 350, With 8 Primes Within It = 139 Repeated Even Nos.
(gg) Percentage Increase In Repetition = (139 - 115) ÷ 115 x 100% = 20.87%

(h) Set Of Integers, 351 To 400, With 8 Primes Within It = 172 Repeated Even Nos.
(hh) Percentage Increase In Repetition = (172 - 139) ÷ 139 x 100% = 23.74%

(i) Set Of Integers, 401 To 450, With 9 Primes Within It = 196 Repeated Even Nos.
(ii) Percentage Increase In Repetition = (196 - 172) ÷ 172 x 100% = 13.95%

(j) Set Of Integers, 451 To 500, With 8 Primes Within It = 220 Repeated Even Nos.
(jj) Percentage Increase In Repetition = (220 - 196) ÷ 196 x 100% = 12.24%

(k) Set Of Integers, 501 To 550, With 6 Primes Within It = 247 Repeated Even Nos.
(kk) Percentage Increase In Repetition = (247 - 220) ÷ 220 x 100% = 12.27%

(l) Set Of Integers, 551 To 600, With 8 Primes Within It = 268 Repeated Even Nos.
(ll) Percentage Increase In Repetition = (268 - 247) ÷ 247 x 100% = 8.5%

(m) Set Of Integers, 601 To 650, With 9 Primes Within It = 298 Repeated Even Nos.
(mm) Percentage Increase In Repetition = (298 - 268) ÷ 268 x 100% = 11.19%

(n) Set Of Integers, 651 To 700, With 7 Primes Within It = 320 Repeated Even Nos.
(nn) Percentage Increase In Repetition = (320 - 298) ÷ 298 x 100% = 7.38%

(o)   Set Of Integers, 701 To 750, With 7 Primes Within It = 340 Repeated Even Nos.
(oo) Percentage Increase In Repetition = (340 - 320) ÷ 320 x 100% = 6.25%

(p)   Set Of Integers, 751 To 800, With 7 Primes Within It = 367 Repeated Even Nos.
(pp) Percentage Increase In Repetition = (367 - 340) ÷ 340 x 100% = 7.94%

(q)   Set Of Integers, 801 To 850, With 7 Primes Within It = 392 Repeated Even Nos.
(qq) Percentage Increase In Repetition = (392 - 367) ÷ 367 x 100% = 6.81%

(r)   Set Of Integers, 851 To 900, With 8 Primes Within It = 412 Repeated Even Nos.
(rr) Percentage Increase In Repetition = (412 - 392) ÷ 392 x 100% = 5.1%

(s)   Set Of Integers, 901 To 950, With 7 Primes Within It = 433 Repeated Even Nos.
(ss) Percentage Increase In Repetition = (433 - 412) ÷ 412 x 100% = 5.1%

(t)   Set Of Integers, 951 To 1,000, With 7 Primes Within It = 470 Repeated Even Nos.
(tt) Percentage Increase In Repetition = (470 - 433) ÷ 433 x 100% = 8.55%

(u)   Set Of Integers, 1,001 To 1,050, With 8 Primes Within It = 492 Repeated Even Nos.
(uu) Percentage Increase In Repetition = (492 - 470) ÷ 470 x 100% = 4.68%

(v)   Set Of Integers, 1,051 To 1,100, With 8 Primes Within It = 523 Repeated Even Nos.
(vv) Percentage Increase In Repetition = (523 - 492) ÷ 492 x 100% = 6.3%

(w)   Set Of Integers, 1,101 To 1,150, With 5 Primes Within It = 545 Repeated Even Nos.
(ww) Percentage Increase In Repetition = (545 - 523) ÷ 523 x 100% = 4.21%

(x)   Set Of Integers, 1,151 To 1,200, With 7 Primes Within It = 553 Repeated Even Nos.
(xx) Percentage Increase In Repetition = (553 - 545) ÷ 545 x 100% = **1.47%**

(y)   Set Of Integers, 1,201 To 1,250, With 8 Primes Within It = **592** Repeated Even Nos.
(yy) Percentage Increase In Repetition = (592 - 553) ÷ 553 x 100% = **7.05%**

It could be seen above that on the whole the No. Of Old/Repeated (Also Appeared Earlier) Even Numbers/Overlaps "Generated" (By The Additions/Combinations Of Two Primes) increases progressively from 20 in (b) to 592 in (y), while it could be seen that the Percentage Increase In Repetition on the whole thins out from 130% in (cc) to 7.05% in (yy), with the lowest percentage increase of 1.47% found in (xx). This statistical trend or feature is not surprising and represents significant evidence that lends support to the possibly validity of the Goldbach conjecture - the infinitude of both the primes and the even numbers implies that the above overlaps increase progressively to infinity.

It is evident here that the higher up the primes we go the more "overwhelmingly" the even numbers "generated" would repeat themselves and overlap. This is significant. Though the infinitude of the prime numbers would ensure that there would always be new even numbers being "generated", there is also the "fear" that there might be gaps, breaks or lack of continuity in the even numbers thus "generated" wherein some of the even numbers in-between can never be sums of 2 primes (as are shown in the example in Figure 2 above), thereby disproving the Goldbach conjecture. But, it is evident that these more and more profuse repetitions and overlaps of the even numbers thus "generated" by the primes the higher up the infinite list of prime numbers we go "ensure" that such gaps or breaks would not appear between the even numbers "generated" - they "ensure" that the even numbers thus "generated" by the primes in the infinite list of primes would be regular, continuous, *without breaks or gaps*, and, in consecutive running order. This evident greater and greater effusiveness or exuberance of the repetitions and overlaps of the even numbers thus "generated" by the primes the higher up the infinite list of prime numbers we go can be likened to a "play-safe measure" wherein there is "safety derived from large numbers". In other words, since an even number could be formed in so many ways by adding 2 primes, i.e., so easily formed by adding 2 primes, evidently more so the higher up the infinite list of prime numbers we go, as has been shown above, the sums of 2 primes thus becoming evidently a defining characteristic of the even numbers, we could expect every larger and larger even number to be the sum of 2 primes in more and more ways as has been shown above (and in the appendices below).

We note again that a long, impressive list of consecutive even numbers, from 4 to 12×10^{17}, had already been verified to be sums of 2 primes, and, these partitions/"prime + prime = even number" combinations would become increasingly more "overwhelming", dense, and repetitive (overlapping) towards infinity (*the Goldbach conjecture becoming evidently stronger and stronger the higher up the infinite list of prime numbers/even numbers we go*), as is described above. The moot question now is, of course, whether after 12×10^{17} there would be an even number in the infinite list of even numbers which is the last, or, largest, even number that is the sum of 2 primes - this largest even number, if it exists (thereby implying the falsehood of the Goldbach conjecture), must (of necessity) be the sum of 2 primes that are each the largest existing prime. However, as the primes are infinite (vide Euclid's theorem), a largest existing prime is an impossibility. Therefore, there can never be a largest even number comprising of the summation of 2 largest existing primes which would disprove the Goldbach conjecture. As a matter of fact, the infinity of the primes implies that there would be an infinite number of double primes which sum up to an even number.

It could thus be concluded that the Goldbach conjecture is possibly valid.

Conclusion

A number of methods have been adopted to show the possibly validity of the Goldbach conjecture.

The inductive method, which is a well-established argument, is one of the methods utilised. The following lends support to this inductive argument supporting the Goldbach conjecture: (a) The characteristic of a mountain or infinite volume of sand is reflected in the characteristic of some grains of sand found there so that studying the characteristic of some grains of sand found there is enough for deducing the characteristic of the mountain or infinite volume of sand, to ascertain the quality of a batch of products it is only necessary to inspect some carefully selected samples from that batch of products and not every one of the products and to carry out a population census, i.e., find out the characteristics of a population, it is only necessary to carry out a survey on some carefully selected respondents and not the whole population; in like manner, by the same principle, we just need to study a carefully selected list of even numbers, find out whether they are all sums of 2 primes and deduce by induction whether all even numbers after this list would also be sums of 2 primes - this act is similar to extrapolation. (For example, a distributed computer search completed in 2008 at the University of Aveiro, Portugal, had confirmed that every even number up to 12×10^{17}, which is no small list of numbers, is the sum of 2 primes. By the principle of induction in this case we could deduce that all the even numbers after 12×10^{17} would also be sums of 2 primes.) (b) Thus, in this way every even number after 2 could be reasonably shown to be the sum of 2 primes. In fact, induction plays an important part in the argument.

The other argument used to support the conjecture is the indirect (reductio ad absurdum) method, which had been used by Euclid and other mathematicians after him. Logically, 1 or 2 examples of "contradiction" should be sufficient evidence of infinity, for it does not make sense to have a need for an infinite number of cases of "contradiction", as our argument would then have to be infinitely and impossibly long, an absurdity. This method of argument is "argument by implication" as a result of "contradiction" - which is a "short-cut" and smart way in showing infinity, instead of "showing infinity by counting to infinity", which is ludicrous, and, impossible. Hence, 1 or 2 cases of "contradiction" should be sufficient for implying that there would be an infinitude of even numbers which are sums of 2 primes, which of course also tacitly implies that there would be an infinitude of the number of cases of such "contradiction". (Euclid evidently had this logical point in mind when he formulated the indirect (reductio ad absurdum) argument for the infinity of the primes.) This method of argument had been cleverly used by a number of mathematicians, not the least by the great German mathematician, David Hilbert. For example, Hilbert had used an indirect method (the "reductio ad absurdum" argument) to argue in support of Gordan's Theorem without having to show an actual "construction", an argument which had been accepted by his peers.

One important query here, which many might not have considered: What if the list of prime numbers is not infinite? Of course, if that is the case, the Goldbach conjecture would be false. It would then have been absurd for the Goldbach conjecture to have been conceived at all. However, the list of primes is infinite (vide Euclid's theorem). This gives credence to the Goldbach conjecture.

A very important related point must be highlighted here. If the Goldbach conjecture were indeed false, there must be an ultimate (largest) even number which is (and must necessarily be) the result of the summation of 2 primes that are each the largest existing prime. It must be noted that this is actually an impossibility, as there can never be a largest existing prime - by Euclid's theorem, the primes are infinite (refer to argument just above). Hence, the Goldbach conjecture has no likelihood of being false, and, by both reductio ad absurdum (contradiction), and, induction (wherein all even numbers up to 12×10^{17}, a long, impressive list, had been confirmed to be sums of 2 primes), is possibly true.

Another very important, perhaps the most important, point is that *the Goldbach conjecture becomes evidently stronger and stronger the higher up the infinite list of prime numbers/even numbers we go* (this is strong, perhaps the strongest, positive indication of the possibly validity of the Goldbach conjecture), as has been shown above in many ways. Thus, by implication, induction, extrapolation, it could be concluded that the Goldbach conjecture is possibly valid - that possibly every even number after 2 is the sum of 2 primes.

So far, there has been no indication or confirmation at all that the number of even numbers after the number 2 which are each the sum of 2 primes is finite and the largest existing even number which is the sum of 2 primes has not been found and confirmed. (This would of course be the case if the Goldbach conjecture is true.) Also, no counter-example (i.e., an even number which is never the sum of 2 primes) has been found so far. On the other hand, practically everyone could intuit that the list of even numbers after the number 2 which are each the sum of 2 primes is infinite. Besides, the evidence, as shown herein, is strongly in support of the possibly infinity of this list.

Appendix 1

(20) <u>Set Of Integers, 1,201 To 1,250, With 8 Primes Within It</u>
    (a) Primes: 1,201; 1,213; 1,217; 1,223; 1,229; 1,231; 1,237 and 1,249
    (b) No. Of Primes: 8
    (c) No. Of Even Numbers "Generated" (Including Repetitions) By The 8 Primes = 648 (1,204 [1,201 + 3] To 2,498 [1,249 + 1,249])
    (d) No. Of New Even Numbers "Generated" = 56 (2,388 To 2,498)
    (e) No. Of Old/Repeated (Also Appeared In (19) Above, With Some Also Having Appeared In (18), (17), (16), (15), (14), (13), (12), (11), (10), (9) And (8) Above) Even Numbers "Generated" (I.e., Repetitions/Overlaps) = 592 (1,204 To 2,386)
    (f) Density Of New Even Numbers "Generated" = (d) ÷ 8 Primes = 56 ÷ 8 = 7 New Even Numbers Per Prime Number

Appendix 2

(8) 8 th. Batch Of 30 Even Numbers (424 To 482) - Partitions/"Prime + Prime = Even Number" Combinations
    (a) 424: No. Of Above-mentioned Prime Additions/Combinations = 12
    (b) 426: No. Of Above-mentioned Prime Additions/Combinations = 21
    (c) 428: No. Of Above-mentioned Prime Additions/Combinations = **9**
    (d) 430: No. Of Above-mentioned Prime Additions/Combinations = 14
    (e) 432: No. Of Above-mentioned Prime Additions/Combinations = 19
    (f) 434: No. Of Above-mentioned Prime Additions/Combinations = 14
    (g) 436: No. Of Above-mentioned Prime Additions/Combinations = 11
    (h) 438: No. Of Above-mentioned Prime Additions/Combinations = 22
    (i) 440: No. Of Above-mentioned Prime Additions/Combinations = 15
    (j) 442: No. Of Above-mentioned Prime Additions/Combinations = 13
    (k) 444: No. Of Above-mentioned Prime Additions/Combinations = 22
    (l) 446: No. Of Above-mentioned Prime Additions/Combinations = 12
    (m) 448: No. Of Above-mentioned Prime Additions/Combinations = 13
    (n) 450: No. Of Above-mentioned Prime Additions/Combinations = 29
    (o) 452: No. Of Above-mentioned Prime Additions/Combinations = 14
    (p) 454: No. Of Above-mentioned Prime Additions/Combinations = 12
    (q) 456: No. Of Above-mentioned Prime Additions/Combinations = 26
    (r) 458: No. Of Above-mentioned Prime Additions/Combinations = **9**
    (s) 460: No. Of Above-mentioned Prime Additions/Combinations = 17
    (t) 462: No. Of Above-mentioned Prime Additions/Combinations = **30**
    (u) 464: No. Of Above-mentioned Prime Additions/Combinations = 13
    (v) 466: No. Of Above-mentioned Prime Additions/Combinations = 14
    (w) 468: No. Of Above-mentioned Prime Additions/Combinations = 26
    (x) 470: No. Of Above-mentioned Prime Additions/Combinations = 16
    (y) 472: No. Of Above-mentioned Prime Additions/Combinations = 14
    (z) 474: No. Of Above-mentioned Prime Additions/Combinations = 24
    (aa) 476: No. Of Above-mentioned Prime Additions/Combinations = 14
    (bb) 478: No. Of Above-mentioned Prime Additions/Combinations = 12
    (cc) 480: No. Of Above-mentioned Prime Additions/Combinations = **30**
    (dd) 482: No. Of Above-mentioned Prime Additions/Combinations = 11
    (i) Maximum No. Of Prime Additions/Combinations = 30
    (ii) Minimum No. Of Prime Additions/Combinations = 9
    (iii) Total No. Of Prime Additions/Combinations For (a) To (dd) = 508
    (iv) Total No. Of Even Numbers = 30
    (v) Density Of Distribution = Average Prime Additions/Combinations Per Even Number = (iii) ÷ (iv) = 508 ÷ 30 = 16.93 Prime Additions/Combinations Per Even Number

9

PRIMES DISTRIBUTION AND RIEMANN HYPOTHESIS

The Riemann hypothesis is an important outstanding problem in number theory as its validity will affirm the manner of the distribution of the prime numbers. It posits that all the non-trivial zeros of the zeta function ζ lie on the critical strip between $\text{Re}(s) = 0$ and $\text{Re}(s) = 1$ at the critical line $\text{Re}(s) = 1/2$. This has been found to be true for the first 10^{13} non-trivial zeros. The locations of these non-trivial zeros on the critical strip are described by a complex number $s = 1/2 + bi$ where the real part is $1/2$ and i represents the square root of -1. It should be noted that the mathematical operations and logic of the complex numbers $a + bi$, where a and b are real numbers and i is the imaginary number square root of -1, are practically the same as for the real numbers and are even more versatile. For the zeta function $\zeta(s)$ to be zero, its series would have to have both the positive terms and negative terms cancelling each other out, though the positive or "+" signs in the series may indicate positive values only which is misleading. The important question is whether there would be zeros appearing at other locations on this critical strip, e.g., at $\text{Re}(s) = 1/4, 1/3, 3/4$, or, $4/5$, etc., which would disprove the Riemann hypothesis.

The role of the non-trivial zeros of the Riemann zeta function ζ, which are mysterious and evidently not much understood, would be explained.

To understand what Riemann wanted to achieve with the non-trivial zeros, we need to understand the part played by the complex plane.

First, the terms in the Riemann zeta function ζ:-

$$\zeta(s) = \sum_{n=1}^{\infty} 1/n^s = 1 + 1/2^s + 1/3^s + 1/4^s + 1/5^s + \ldots$$

where s is the complex number $1/2 + bi$

For the term $1/2^{1/2 + bi}$ above, e.g., whether it would be positive or negative in value would depend on which part of the complex plane this term $1/2^{1/2 + bi}$ would be found in, which depends on $2(n)$ and b (it does not depend on $1/2 - 1/2$ and $2(n)$ only determine how far the term is from zero in the complex plane). This term could be in the positive half (wherein the term would have a positive value) or the negative half (wherein the term would have a negative value) of the complex plane. Thus, some of the terms in the Riemann zeta function ζ would have positive values while the rest have negative values (depending on the values of n and b). The sum of the series in the Riemann zeta function ζ is calculated with a formula, e.g., the Riemann-Siegel formula, or, the Euler-Maclaurin summation formula.

Riemann evidently anticipated that there would be an equal, or, almost equal number of primes among the terms in the positive half and the negative half of the complex plane when there is a zero. In other words, he thought that the distribution of the primes would be statistically fair, the more terms are added to the Riemann zeta function ζ, the fairer or "more equal" would be the distribution of the primes in the positive half and the negative half of the complex plane when there is a zero. (Compare: The tossing of a coin wherein the more tosses there are the "more equal" would be the number of heads and the number of tails.) That is, in the longer term, with more and more terms added to the Riemann zeta function ζ, more or less 50% of the primes should be found in the positive half of the complex plane and the balance 50% should be found in the negative half of the complex plane, the more terms there are the fairer or "more equal" would be this distribution, when there is a zero.

A non-trivial zero indicates the point in the Riemann zeta function ζ wherein the total value of the positive terms equals the total value of the negative terms. There is an infinitude of such points, i.e., non-trivial zeros. Riemann evidently thought that for the case of a zero the number of primes found among the positive terms would be more or less equal to the number of primes found among the negative terms, which represents statistical fairness. It is evident that through a zero the order or pattern of the distribution of the primes could be observed.

Next, the error term in the following J function for calculating the number of primes less than a given quantity:-

$$J(n) = Li(n) - \sum_p Li(n^p) - \log 2 + \int_n^\infty dt/(t(t^2 - 1) \log t)$$

where the 1st. term $Li(n)$ is generally referred to as the "principal term" and the 2nd. term $\sum_p Li(n^p)$ had been called the "periodic terms" by Riemann, Li being the logarithmic integral

$\sum_p Li(n^p)$, the secondary term of the function, the error term, represents the sum taken over all

the non-trivial zeros of the Riemann zeta function ζ. n here is a real number raised to the power of p, which is in this instance a complex number of the form $1/2 + bi$, for some real number b, $n^{1/2}$ being $\sqrt{n}$. If the Riemann hypothesis is true, for a given number n, when computing the values of n^p for a number of different zeta zeros p, the numbers we obtain are scattered round the circumference of a circle of radius $\sqrt{n}$ in the complex plane, centred on zero, and are either in the positive half or negative half of the complex plane.

To evaluate $\sum_p Li(n^p)$ each zeta zero has to be paired with its mirror image, i.e., complex conjugate,

in the south half of the argument plane. These pairs have to be taken in ascending order of the positive imaginary parts as follows:-

$$\text{zeta zero: } 1/2 + 14.134725i \ \& \ \text{ its complex conjugate: } 1/2 - 14.134725i$$
$$\text{zeta zero: } 1/2 + 21.022040i \ \& \ \text{ its complex conjugate: } 1/2 - 21.022040i$$
$$\text{zeta zero: } 1/2 + 25.010858i \ \& \ \text{ its complex conjugate: } 1/2 - 25.010858i$$
$$\cdot$$
$$\cdot$$
$$\cdot$$

(<u>Note</u>: The complex conjugates are all also zeros.)

If, e.g., we let $n = 100$, then the error term for $n = 100$ would be $\sum_p Li(100^p)$. To calculate this

error term, we have to first raise 100 to the power of a long list of zeta zeros in ascending order of the positive imaginary parts (the 1st. 3 zeta zeros are shown above), the longer the list of zeta zeros the better, e.g., 100,000 zeta zeros, in order to achieve the highest possible accuracy in the error term. Then we take the logarithmic integrals of the above powers (100,000 pairs of zeta zeros & their complex conjugates) and add them up, which is as follows:-

$$100^{1/2 + 14.134725i} + 100^{1/2 - 14.134725i}$$

$$+ 100^{1/2 + 21.022040i} + 100^{1/2 - 21.022040i}$$
$$+ 100^{1/2 + 25.010858i} + 100^{1/2 - 25.010858i}$$

.

.

.

The imaginary parts of the zeta zeros would cancel out the imaginary parts of their complex conjugates, leaving behind their respective real parts. For example, for the 1st. zeta zero $1/2 + 14.134725i$, its imaginary part $+14.134725i$ would cancel out the imaginary part $-14.134725i$ in its complex conjugate $1/2 - 14.134725i$, leaving behind only the real parts $100^{1/2}$ for each of them. That is, for $100^{1/2 + 14.134725i} + 100^{1/2 - 14.134725i}$, we only have to add together the logarithmic integral of $100^{1/2}$(from $100^{1/2 + 14.134725i}$) and the logarithmic integral of $100^{1/2}$(from $100^{1/2 - 14.134725i}$) to get the 1st. term. The same is to be carried out for the next 99,999 powers in ascending order of the positive imaginary parts, giving altogether a total of 200,000 logarithmic integrals (of both the zeta zeros & their complex conjugates) to be added together to give the 100,000 terms. These terms have either positive or negative values, an equal or almost equal number of positive and negative values, which depend on whether they are in the positive or negative half of the complex plane, as is described above. The positive values and the negative values of these 100,000 terms are added together and should cancel out each other, slowly converging. The difference between the positive values and the negative values of these 100,000 terms constitutes the error term. (Note that the Riemann hypothesis asserts that the difference between the true number of primes $p(n)$ and the estimated number of primes $q(n)$ would be not much larger than $\sqrt{n}$ - not much larger than $\sqrt{100}$ ($\sqrt{100}$ is also expressed as $100^{1/2}$) in the above case. Like the case of tossing a coin wherein the statistical probability is that in the long run the number of heads would practically equal the number of tails, there should be equal or almost equal quantities of positive terms and negative terms, i.e., 50,000 or thereabout positive terms and 50,000 or thereabout negative terms, which would be statistically fair, the discrepancy if any being the error.)

All this is evidently a laborious process, though the ingenuity of the ideas behind the Riemann hypothesis should be acknowledged.

It may be compared to the sieve of Eratosthenes, which could find the exact number of primes less than a given quantity without any error at all.

FURTHER READING

[1] D. Burton, Elementary Number Theory, Allyn & Bacon, 1980
[2] R. Courant and H. Robbins, revised by I. Stewart, What Is Mathematics? An Elementary Approach to Ideas and Methods, Oxford University Press, 1996
[3] R. L. Devaney, An Introduction to Chaotic Dynamical Systems, Benjamin-Cummings, 1986
[4] P. Embrechts, C. Kluppelburg, and T. Mikosch, Modelling Extremal Events for Insurance and Finance, Berlin: Springer-Verlag, 1997
[5] P. Embrechts, Where Mathematics, Insurance and Finance Meet, Quantitative Finance, 2 (2002) (6): 402-404
[6] G. H. Hardy and E. M. Wright, An Introduction To Theory Of Numbers, Oxford, England: Clarendon Press, 1979
[7] D. H. Lehmer, List Of Prime Numbers From 1 To 10,006,721, Publication No. 65, Carnegie Institution of Washington, Washington, D.C., 1914
[8] M. E. Lines, A Number For Your Thoughts, Adam Hilger, 1986
[9] B. B. Mandelbrot, The Fractal Geometry Of Nature, W. H. Freeman, 1977
[10] J. M. T. Thompson & H. B. Stewart, Nonlinear Dynamics and Chaos, John Wiley, 1986
[11] S. Vaseghi, The Secret Harmony of Primes, Chiron Academic Press, 2016
[12] R. Wilson, Number Theory: A Very Short Introduction, Oxford University Press, 2020

www.ingramcontent.com/pod-product-compliance
Lightning Source LLC
Chambersburg PA
CBHW082115220526
45472CB00009B/2185